Eliminating Engine Interference

by JOHN D. LENK

HOWARD W. SAMS & CO., INC.
THE BOBBS-MERRILL CO., INC.
INDIANAPOLIS · KANSAS CITY · NEW YORK

Preface

As an amateur or CB operator, have you ever found it necessary to stop your engine so that you could complete a radiotelephone conversation? As a service technician, do your customers complain of so much noise on their mobile two-way radio that they can't tell one signal from another? When operating marine electronic equipment do you notice a continuous set of traces or flashes on a depth sounder that seem to mask the true depth readings? If your answer is yes to any of these questions, you probably have radio interference caused by the ignition system of the engine.

Interference has long been one of the most annoying factors in mobile and marine communications systems. Equipment manufacturers have taken what steps they can to combat this noise problem by adding shielding and electronic filters to the sensitive portions of the receivers. Most have incorporated squelch and automatic noise-limiter circuits. Effective as these circuits are, they only set the noise level and make it bearable. Weak signals that are below this level simply do not get through. The result is a limiting of the operating range of a receiver and a possible loss of communication.

This book is written to provide a practical approach to the problem of engine interference. All the information is based on the assumption that the man in the field is interested in two facts: how to pinpoint the source of the trouble and how to eliminate it.

Many companies are working on devices to reduce the amount of interference created by engines—both automotive and marine. Manufacturers of mobile and marine electronic equipment are trying to devise circuits that will permit their equipment to function satisfactorily in spite of electronic noise. I have tried to combine the information available from both groups so that you can get the maximum benefit. Engine interference can be a difficult and annoying problem, but it can be licked.

JOHN D. LENK

Contents

CHAPTER 1

CHAPTER 2

CHAPTER 3

CHAPTER 4

CHAPTER 5

CHAPTER 6

CHAPTER 7

A Look at Interference

Interference, created by the electrical system of an engine, has been a problem for a long time. Both radio and audio frequencies may be involved, though the former are the more serious. Current trends in engine design and increased receiver sensitivity tend to aggravate the situation instead of helping it. Actually, it does not make any difference whether the engine is in an automobile or a boat; it can still create a serious interference or noise problem.

The major sources of engine interference are related to essential parts of the motor, so there is no hope of eliminating the problem completely. With a clear understanding of the principles involved, however, it is possible to pinpoint the exact causes of the interference. Steps can then be taken so that communications are disrupted as little as possible. Providing the necessary information is the purpose of this book.

DEFINITIONS

Interference is any electrical disturbance that causes an undesirable response or a malfunctioning of communications receivers or other electronic equipment. Noise in a receiver is the most common form, but it may also show up as misleading readings on a depth finder or as other undesirable signals. In this chapter the discussion is confined to electrical interference caused by engines.

Engine interference can be broken down into two classes or types according to the method by which the interference reaches the receiver. Radiated interference is unwanted electrical energy broadcast into the area surrounding the spark plugs, the distributor, the ignition coil, and the spark-plug wires themselves. This energy is similar in characteristics to radio waves or magnetic fields. Normally it is picked up by the receiver antenna. Conducted interference is noise carried through the electrical wiring or conducted into the receiver via the wiring.

It is not always easy to distinguish between the types of interference; quite often the two are mixed. Any wire carrying conducted electrical interference can serve as an antenna and radiate this interference. Likewise, nearby wires can pick up radiated electrical interference and conduct it just as though they were physically connected to the line.

CAUSES

All engine electrical noise or interference originates from the same source—sparking. If you are familiar with the early days of radio, you know that the spark-gap principle was used in radio transmitters. Radio waves were created when an electrical spark jumped across a gap; these were picked up by the receiver. By carefully timing the duration of the spark and the interval between sparks, a code was transmitted without the use of wires between the transmitter and the receiver.

The gasoline engine is full of spark-noise sources. Spark plugs, distributor points, distributor contacts, generator brushes, and voltage-regulator contacts all have gaps that electricity must jump in the normal operation of the motor. These parts are divided among three circuits. First is the high-voltage secondary circuit, consisting of spark plugs, distributor contacts, and ignition-coil secondary winding (Fig. 1-1). This circuit is the source of the worst radio interference.

Next is the low-voltage primary ignition circuit that consists of the distributor points, condenser, and ignition-coil primary. Finally, there is the generator and voltage-regulator circuit.

To best understand how these circuits create radio noise or interference, you need to review how these circuits operate. The distributor points are opened and closed by a cam at a rate determined by the speed of the engine. When the points are closed, there is current from the battery through the ignition switch, distributor points, and primary winding of the ignition coil. There is no current in the secondary circuit at this time. The magnetic field collapses when the points open, inducing a voltage in the secondary winding of the ignition coil.

The secondary winding of the coil has many more turns than the primary—usually a ratio of from 50:1 to 100:1. The sudden collapse of the magnetic field when the distributor points open induces a voltage across the coil secondary that is proportional to the number of turns on the secondary winding. The result is a force of over 10,000 volts that discharges across contacts in the distributor cap

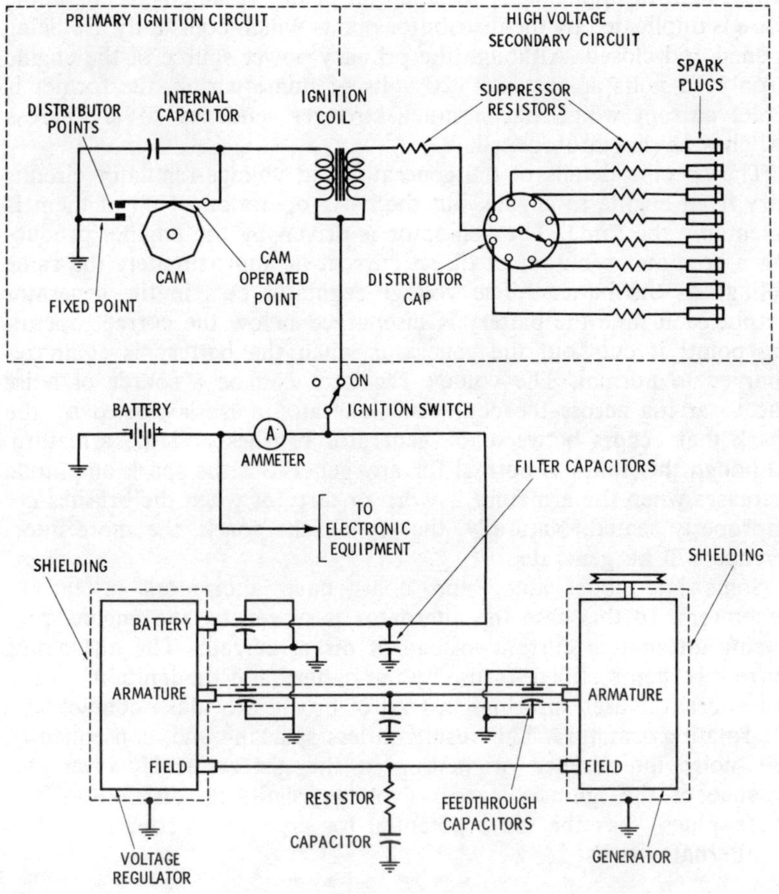

Fig. 1-1. Typical automobile electrical system.

and one spark plug. Consequently, there are two high-voltage arcs or sparks each time the distributor points open—one at the spark plug to fire the gasoline and the other in the distributor cap.

Radio waves are created whenever a spark jumps between two contacts. In the engine, radio waves are generated in the secondary ignition circuit. They are picked up by any electronic gear and appear as noise or interference in the speaker.

The primary ignition circuit also produces noise because there is a spark when the breaker points start to open. You have probably noticed that a noise (or pop) is produced in your home radio each time a light or appliance is switched off or on. In an engine this same noise is duplicated by the distributor points which constantly are being opened and closed. Although the primary power source of the engine is only 12 volts against the 120 volts of home wiring, the former is direct current which has a much stronger tendency to arc across switches as they are opened.

The specific details of the generator and voltage-regulator circuits vary from engine to engine, but the basic operation of all of them is essentially the same. The generator is driven by the engine, producing a continuous output of direct current at approximately the same voltage as the battery. The voltage regulator cuts in the generator output each time the battery is discharged below the correct operating point; it cuts out the generator when the battery is again recharged to normal. The voltage regulator can be a source of noise due to arcing across the contacts. Generator noise is caused by the spark that occurs between the generator brushes and the armature. Although this spark is normal for any generator, the spark amplitude increases when the armature is worn or dirty or when the brushes are improperly seated. Naturally, the greater the spark, the more interference will be generated.

Some late-model autos and boats have alternators instead of generators. In this case the alternator is driven by the engine, producing alternating current instead of direct current. The alternating current is then rectified for use by the battery and the ignition system. An alternator uses slip-rings instead of brushes to make contact with the rotating armature. This results in less sparking and, consequently, less noise interference originating in the generator. However, the balance of the ignition system (voltage regulator, distributor, and spark plugs) has the same potential for creating interference when an alternator is used.

DEFECTIVE IGNITION SYSTEMS

Well-adjusted and carefully-tuned engines can create some radio interference; a defective ignition system can really add to the problem. If spark plugs, distributor points, and distributor contacts are fouled, worn, or defective, there will be more sparking or a longer spark that will create more noise. The same is true if generator brushes are dirty, if the spark-plug gap has widened, if alternator slip rings are dirty, or if the voltage-regulator points are not making good contact. Notice that these conditions can exist without affecting the operation of the engine enough to be apparent to the average driver. They should be checked and corrected before the installation of a noise-suppression system. However, it is possible for a perfectly maintained engine to produce enough electrical noise to interfere seriously with mobile communications.

TRENDS IN ENGINE DESIGN

There are several reasons why interference is becoming an increasing problem:

1. Receivers in all communications frequencies have been improved and are more sensitive than five years ago. This increasing sensitivity means the receivers are able to transform weaker signals into meaningful sound. Low-level interference consequently becomes more noticeable and more objectionable.
2. In the same period of time, horsepower and displacement of engines have increased tremendously (Table 1-1). Higher voltages for spark plugs have become the rule. Increased power has resulted in greater wear on critical parts so electrical interference has increased.

Table 1-1. Comparison of Automobile Engines.

	1960	1965*
Maximum Compression Ratio	8:1	11:1
Cubic Inch Displacement	348	425
Maximum Brake Horsepower	250	350

*Average 8-cylinder passenger vehicle engine

3. Changes in body and frame construction through the use of plastics and composition materials have introduced electrical problems. It is difficult to provide good, effective bonding and grounding when nonconducting materials are used.
4. Narrow-band or split-channel operation (by FCC regulation) reduces the strength of the received signal in the low f-m band. This problem can develop on other frequencies in the future because of the demand for spectrum space.

EFFECT ON COMMUNICATION

Ignition noise is not only audibly offensive; it limits the effectiveness of communications. Basically, all gasoline engines create noise although the amplitude and characteristics will vary from vehicle to vehicle. The fact remains that noise is there and travels with the receiver. Signal strength depends on the type, power, and frequency of operation of the mobile and base-station equipment. It is also affected by topography, base-station antenna height, building heights, skip effect, and a variety of other factors. Mobile equipment will receive a given transmission adequately without disruption from noise as long as the signal strength is above a certain level. As soon as the vehicle moves out of this range, or the transmission is weakened by any one of the variable factors, electrical noise being radiated or conducted will become relatively stronger in intensity and eventually override the strongest of signals, thus limiting the range of communication. Proper noise-elimination techniques can extend the range by removing the noise before it can affect the incoming signal. Sometimes the distance over which communication is possible can be doubled.

AUTOMATIC NOISE LIMITERS AND SQUELCH CIRCUITS

The automatic noise limiters and squelch circuits incorporated into most present-day receivers cannot handle all engine ignition noise. These circuits do a fairly comprehensive job within their design capabilities. However, there is no system offered today that can adequately handle the radio interference from a fouled spark plug or a worn ignition system.

Noise limiters are basically filters to remove certain frequencies that are normally generated by the ignition system. This is effective as long as the interference stays within the prescribed range. In a poorly adjusted motor, however, the noise frequencies range across the entire spectrum so they cannot all be filtered out. In addition, noise such as generator whine is not affected by noise limiters. The manufacturers of much communications equipment state that it is advisable to use bonding, grounding, and suppression kits in conjunction with their product even if noise-limiting circuitry is provided. The cost of adding special noise-limiter equipment to an existing receiver is usually high. Eliminating noise at the source is the positive way to control the problem.

Tomorrow's transistor ignition systems will probably increase, rather than diminish, the problem. With them go hotter sparks, wider gaps, and higher voltages in general, all creating more and louder interference.

EXTERNAL SOURCES OF INTERFERENCE

There are many sources of interference outside of the engine. These range from power lines through arc welders operating in a plant along the side of the road. It is true that effective automatic noise limiters and squelch circuits will reduce the amount of audible noise under such conditions. In heavy traffic interference originating in other engines can become a problem. While engine-noise eliminating techniques handle only your own vehicle, this represents 90% of the problem. This noise is always with you; other noise sources are generally passed or left behind. The problem of external noise is greatest from the marine band (2 to 3 mc) through the 50-mc range. This, of course, includes the 27-mc Citizens-band frequencies where the problem is compounded due to low power operation.

ELIMINATING SPARK NOISE

By now it is obvious that electric sparks from any source are the chief creators of noise interference. These start undesirable radio waves that can be radiated or conducted into electronic gear. Obviously, sparking can be eliminated by stopping the engine or cutting off the generator each time the electronic equipment is used, but this is not a practical answer. The problem must be attacked at its source —the engine itself. Three methods can be used: arc suppression, filtering, and shielding.

Arc Suppression at Spark Plugs

Arc suppression systems are based on three facts: (1) Although it takes a high voltage to make a spark jump across the gap of a spark plug, very little current is needed. (2) The amount of radiated interference is proportional to the current. (3) A high resistance placed in series with the spark plug and the distributor will limit the current to practically nothing without lowering the voltage. Thus, a high resistance in the spark circuit does not affect the spark for firing the engine, but it does reduce the radiated noise in proportion to the reduction of current. In actual practice this is accomplished by installing a suppressor on each spark plug. One is also added in the coil circuit at the distributor.

A suppressor consists of a resistor encased in a tubular insulated housing. It can be added to the engine wiring with very little effort. You only need to lift the ignition lead from the spark plug, place the suppressor on the spark plug, and connect the ignition lead to the suppressor. To install a distributor suppressor, pull the center wire out of its socket in the distributor cap and insert a suppressor. Then plug the wire into the socket in the suppressor. It's that simple.

There is one drawback in using suppressor resistors, even though they are effective. If sufficient resistance is added to a circuit (by means of suppressors) to eliminate all objectionable interference, current may be lowered to the point where the timing or ignition of the engine may be impaired. Yet the maximum that can be added may not be enough to eliminate all objectionable interference. It is quite possible that filtering (for conducted noise), and shielding (for radiated noise) will be required.

Interference Filters for Ignition Systems

The three basic types of interference filters in use today are filter condensers (or capacitors), radio-frequency choke coils, and wave traps. Filter condensers are the most common and, except in special cases, the most effective.

A capacitor will filter out electrical interference because of its reaction to the presence of electrical currents. A capacitor of any type has the property of passing alternating current and fluctuating direct current, but it will not pass pure direct current. If a capacitor is connected between an electrical line and ground, it will provide a low-resistance path to ground for alternating current and fluctuating direct current. Since these are the types of currents produced by the electric sparks, all interference will be routed directly to ground, while the pure direct current will be passed to the battery and power wiring in the normal manner.

The three most common types of filters are shown in Fig. 1-2. Usually these filters capacitors are in a tubular case and have a capacitance value of 0.1 to 0.25 microfarad. In the noninsulated type (Fig. 1-2A) the protruding wire is for connection to the hot side of the circuit, and the mounting lug is fastened to a ground screw on the generator, regulator, engine block, etc. To be effective the mounting lug must make firm contact with clean metal (paint or grease removed) and the head of the mounting bolt. The insulated type of filter capacitor (Fig. 1-2B) has two leads: one for ground and one for the hot lead. It needs to be mounted rigidly. This type of filter capacitor has lost popularity in mobile systems because it lacks mechanical strength.

A feedthrough capacitor (Fig. 1-2C) is generally used in conjunction with shielding where it is necessary to pass a d-c line out of a shielded area. For example, one could be used at a voltage

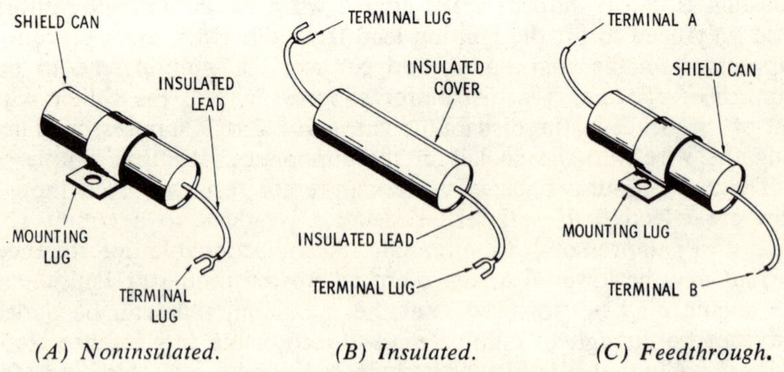

(A) Noninsulated.   *(B) Insulated.*   *(C) Feedthrough.*

Fig. 1-2. Typical filter capacitors.

regulator that has a metal shield. The lead to the battery terminal of the voltage regulator could be removed and connected to terminal A of the condenser. Terminal B would then be connected to the voltage-regulator battery terminal. Then, the current would flow through the capacitor from A to B via a conductor that is internally connected to the ungrounded electrode of the capacitor. The grounded electrode is connected to the capacitor shield that is grounded in turn to the voltage-regulator cover.

Normally a filter capacitor should not be connected directly to the field terminal of a generator or voltage regulator since this could cause damage to the voltage-regulator contacts. (There will usually be a tag or decal on the generator to this effect.) However, a resistor and a capacitor in series can be used on the field terminal as shown in Fig. 1-1. The resistor value should be 4 to 7 ohms; the capacitor value should be about 200 pf.

An r-f choke coil has the opposite effect of a capacitor. It restricts alternating current and fluctuating direct current, while it passes pure direct current. Consequently, choke coils are connected in series with the electrical line they are supposed to protect. They are not frequently used for interference suppression because the commercial chokes used in radio and TV are not usually designed to handle the heavy currents in engine use. Quite often chokes can be made up by winding coils of heavy wire on an experimental basis. These may prove useful in some cases, but it is pretty much of a hit or miss proposition.

Wave Traps

Wave traps are used primarily to filter generator spark interference. They are more difficult to install than filter capacitors and require adjustment after installation (unless they are pretuned at the factory). Their operation is quite simple to undersand. Wave traps are nothing but a coil and capacitor combination (tank circuit) that is tuned to either the frequency of the spark interference or to the operating frequency of the electronic gear. They act to reject or trap interference of this frequency just like wave traps in antenna systems. Their one major drawback is that they can be tuned to only one frequency although spark noise is often generated on several frequencies at once. They can be effective in the case of extreme generator interference on a few channels.

SHIELDING SYSTEMS

Even when suppressors and filters are used, it is still possible for an engine to create sufficient interference to seriously affect mobile or shipboard communications. Filtering is most effective on conducted noise (noise which is conducted from the ignition system or generator to the receiver through the wiring). Radiated noise is a greater problem. As stated previously, signals are generated when there is a spark across a gap. These signals are conducted through the lines in which the spark occurs and are radiated from these lines. Just as in the case of conventional radio waves, electrical interference can be prevented from radiating by surrounding the source of interference with a metal shield. Actually there would be far more ignition interference on all mobile equipment were it not for the fact that the metal firewall and body form a natural shield. The noise problem is more serious in marine engines because this shielding is lacking. Even with automobile installations, however, the radiated interference is picked up by power wiring inside the engine compartment. From there it is conducted through the firewall to the receiver. The only sure way to prevent radiated interference is to surround the source and block the radiations. That is what mechanical shielding systems try to do.

Virtually all aircraft flying today, either piston or turbine powered, utilize shielded ignition systems. Wthout them the elaborate communications and navigation equipment they contain would be useless. Shielding an ignition system provides the only sure means of reducing ignition interference levels. As you might suspect, shielding is the most expensive means of noise elimination. Radio-interference energy that would normally radiate into space from the ignition system is trapped and bottled up by the shielding. This bottling up is accomplished by enclosing all interference-producing components of the ignition system within electrically conductive materials. All shielding components are electrically interconnected and grounded to the engine ground system by very low-resistance (low-impedance) cable. Any high-resistance connections will leak interference.

Shielding may be complete or partial. A complete kit consists of metal shields for the spark plugs, the high-voltage wiring (from the plug to the distributor), the distributor cap, the ignition coil, and possibly, the voltage regulator. Complete systems are higher in cost than the partial shielding systems since it is necessary for the manufacturer to design parts that will fit every make and model of engine.

A typical partial shielding kit consists of spark-plug shields, several feet of shielding braid, a distributor-cap shield, and shield for the ignition coil (with feedthrough filter capacitors.) These kits are sold with complete step-by-step instructions for installation.

Besides these two extremes in shielding systems there are in-between installations where some of the components are assembled while other components must be fabricated on the spot. In addition, there are many do-it-yourself systems described in the various mobile communications publications. Because of the importance of shielding, all of Chapter 7 is devoted to this subject.

IDENTIFYING INTERFERENCE SOURCES

It is possible to identify the source of interference by the sound that comes from a receiver. Here are the characteristics. Ignition interference is a popping sound that is synchronized with the speed of the engine. Generator noise can generally be identified as a whine that starts only when the engine is speeded up. If there is doubt as to the noise source, temporarily remove the generator leads (or slip the generator belt off) and check the noise level with the engine running. If the noise is still there, it originates in the ignition circuit. If the noise is eliminated by disconnecting the generator, the interference originates in the generator.

Instrument noise is identified as hissing or crackling sounds. Instruments can be disconnected one at a time until the noise maker is found. Tire and wheel noises usually result in an irregular rushing sound in the receiver that is present only when the auto is in motion (with or without the engine). A voltage regulator will usually produce a rough, rasping sound.

PRELIMINARY TESTING

The easiest way to make a noise or interference test on any engine is to check the noise level of a receiver with the engine running, then check it under identical conditions with the engine off. Try to make the test away from any external noise source such as other engines, high-voltage lines, neon lights, etc. Turn off the squelch and any other noise-limiter circuits. Adjust the receiver gain until you hear a steady noise level on the speaker. If there are any signals on the air, select a weak signal and adjust the gain until it is barely audible. Then turn off the engine and see if the noise level drops noticeably.

If there is no great change in background noise when the engine is turned off, you do not have an engine interference problem. All is well. Forget the whole thing and count yourself among the fortunate. If you do note any appreciable change, you may as well start looking for the source of the interference. That is what the next six chapters are about.

CHAPTER 2

Basic Ignition-Circuit Suppression

Two different types of noise have to be dealt with in suppressing ignition interference: noise radiated directly from the ignition system, and noise distributed through the entire auto by the electrical wiring and mechanical members (frame, fenders, hood, etc.). Radiation is reduced by either inserting resistors in the ignition wiring or using resistance-type wire. Distributed or conducted noise is reduced with bypass capacitors at various points in the electrical system. Copper braid is used to ground the mechanical members and prevent them from re-radiating the noise.

SUPPRESSOR RESISTORS

Various forms of resistors are used to suppress ignition noise. The resistor shown in Fig. 2-1 is inserted in the ignition lead; the lead is

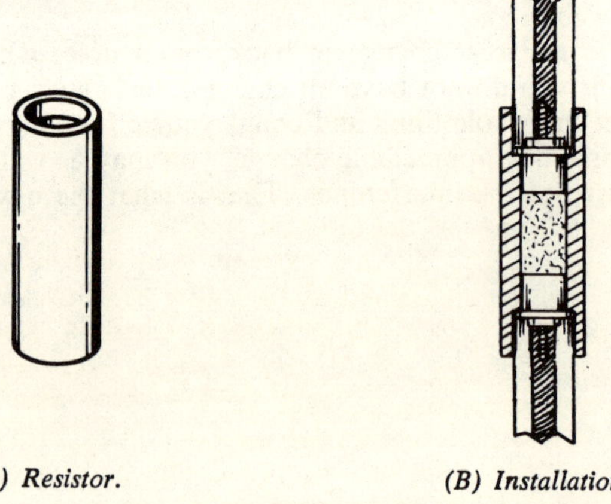

(A) Resistor.   (B) Installation.

Fig. 2-1. Suppressor resistor.

cut in two, and the ends are twisted into the resistor. Suppressors that mount on the spark plug are also available.

The distributor lead connecting the coil to the rotor of the distributor radiates the most noise. In many installations adding suppression to this one lead eliminates most ignition interference. Fig. 2-2 shows a suppressor installed in a distributor lead of a typical auto. When a suppressor is used, it should be installed as close to the distributor as possible, since this is where the sparking takes place. Note that some autos have suppressors built into the distributor cap. In this case it is probably not desirable to increase the resistance.

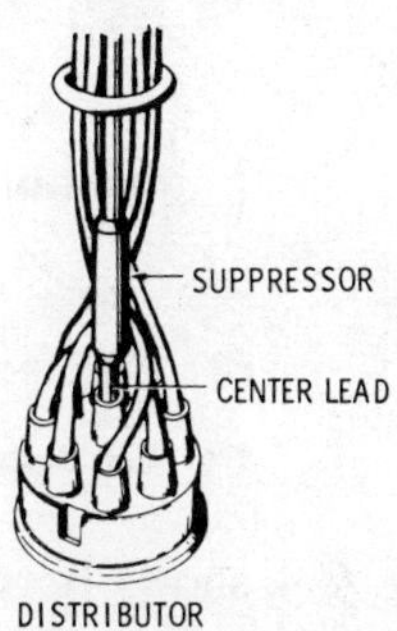

Fig. 2-2. Suppressor installed in the distributor lead.

SUPPRESSOR-TYPE SPARK PLUGS

Suppressor-type spark plugs contain a built-in suppressor resistor. This often eliminates the need for an external suppressor. There are three reasons why resistor spark plugs suppress interference (Fig. 2-3):

1. Interference radiating from cables is greatly reduced in strength.
2. The high-frequency part of the spark is chopped off.
3. The resistor is nearer to the spark gap than would be possible with an external suppressor.

The length of time a suppressor-type spark plug is effective depends on the stability of the internal resistor. Such resistors must be specifically designed to tolerate extreme heat and high voltage. The stability of a resistor is defined as its ability to keep within an ohmic range despite heat and electrical stress. Unstable resistors usually give poor noise suppression. Poorly designed resistors may change in resistance value during normal service; even the best resistors may change in resistance if they are operated at excessive temperature and/or voltage. It should also be noted that excessively high resistance can result in faulty engine performance and poor fuel economy.

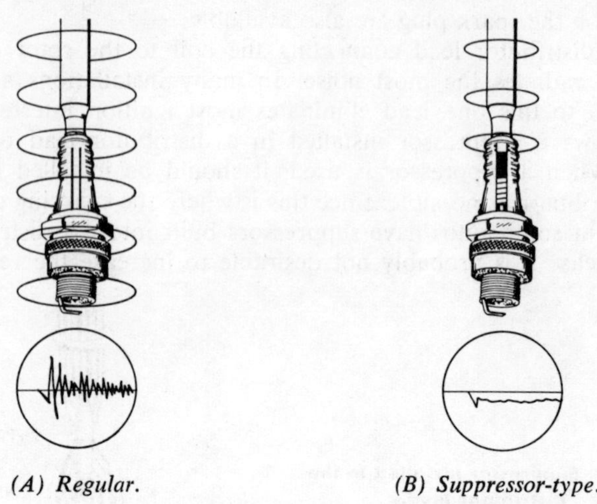

Fig. 2-3. Regular versus suppressor-type spark plugs.

Fig. 2-4 shows how the resistance of suppressor-type spark plugs varies under normal operating conditions. Standard test equipment will not accurately measure spark-plug resistance since heat and high-voltage stress cannot be duplicated out of the engine. For example, it would be impractical to measure a spark-plug resistor at a temperature of 500°F with 5000 volts across it without special equipment. However, a standard ohmmeter can be used for a quick check of the plug resistor. If a resistor in a cold plug exceeds 18,000 ohms on a standard ohmmeter, it should be considered suspect. Of course, if all of the plugs can be checked, each can be compared with the average. In any event, an open-circuit reading on a resistor-type plug indicates a completely failed resistor; that plug must be replaced.

There is no point in making resistance measurements with an ohmmeter on regular or auxiliary-gap plugs. Auxiliary-gap plugs always show an open circuit unless they are defective, and it is possible for a non-resistor plug to show an open circuit without being defective (Fig. 2-5).

SUPPRESSOR CABLE

Most late-model vehicles use some type of resistance wire in the ignition system. This high-voltage ignition cable uses a resistive center conductor instead of low-resistance wire. Several variations have been manufactured in the past, identified as "radio" or "radio resistance"

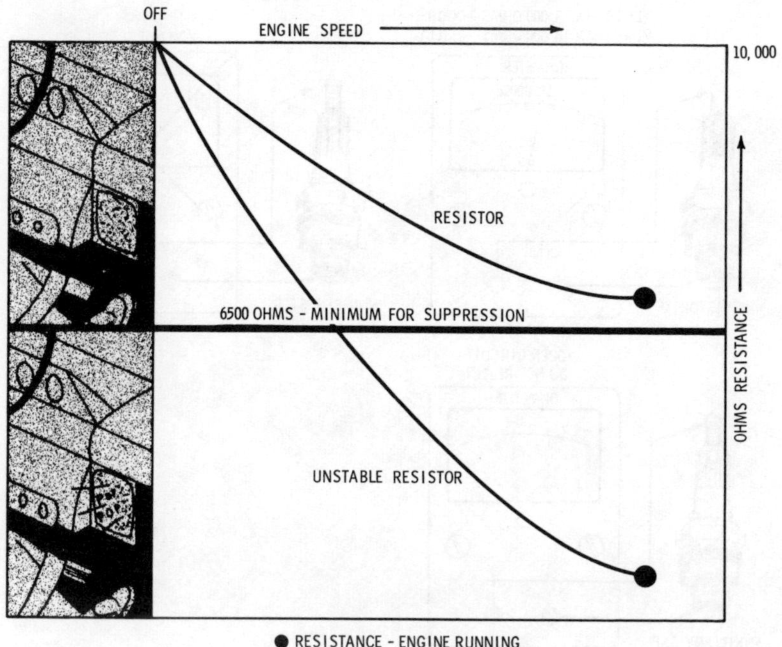

Fig. 2-4. Resistor stability characteristics.

wire. Resistance was usually about 4000 ohms per foot. Present-day standards specify two types: HTLR (3000 to 7000 ohms per foot) and HTHR (6000 to 12,000 ohms per foot).

Resistance values of this cable will generally change during service. Time, temperature, high voltage, vibration, and rough handling take their toll. In general it is good practice to make a resistance check when the ignition system is serviced or if a particular noise problem has been encountered. A rule-of-thumb to follow is this: Replace the cables when the resistance value exceeds 18,000 ohms per foot or falls below 3000 ohms per foot. Three points to remember concerning suppressor-type cable are:

1. Handle resistor cable carefully. Pull on the boot—not on the cable.
2. Never cut a resistor cable to attach a suppressor. The latter should not be used with resistor cable.
3. Never attempt to repair an end terminal. Replace the cable or, preferably, the entire set of cables.

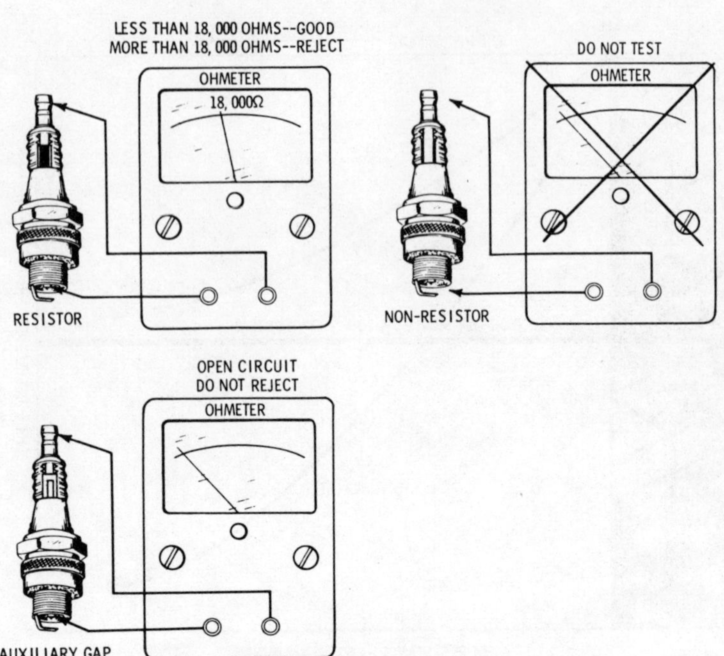

Fig. 2-5. Ohmmeter checking of spark plugs.

LIMITS OF IGNITION-SYSTEM SUPPRESSION

The condition of the ignition system is a major factor in applying suppression. Systems in poor condition cannot tolerate much suppression without causing a noticeable drop in engine performance. However, suppression has no effect on an ignition system in good condition. In most cases resistors are not the cause of poor engine performance. Extensive investigation has revealed that carburetor maladjustment or poor ignition—not unstable spark-plug resistors—is the primary cause of poor performance. In fact, motors usually start faster in subzero weather with resistor-type spark plugs than with standard plugs.

The amount of suppression an ignition system can take depends on:

1. The design of the ignition system.
2. How often the ignition system is serviced.
3. The stability of the resistors (either in the plugs or in external suppressors) used for suppression.
4. The type of vehicle operation. Start-stop driving or long periods of idling increases the tendency for the plugs to foul and

consequently limits the amount of suppression that can be used.

If spark-plug fouling is a problem, adding suppression beyond that supplied with the vehicle may result in poor ignition performance. Consequently, do not add suppressors until the spark-plug problem has been solved.

SUPPRESSORS, RESISTOR PLUGS, AND RESISTANCE WIRE

Two suppression methods are available for any particular system. There is usually a choice of resistor plugs with suppression cable, or resistor plugs with a 10,000-ohm resistor suppressor in the center of the distributor and 5000-ohm suppressors in the towers

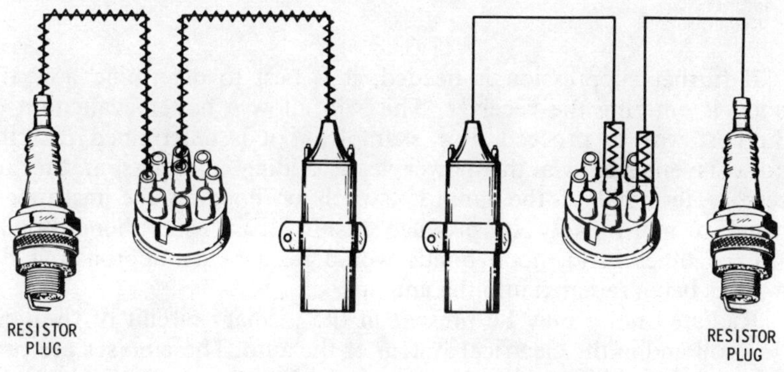

(A) Suppressor-type cable.   *(B) Resistor suppressors.*

Fig. 2-6. Methods for using suppressors.

(Fig. 2-6). Many operators of heavy-duty vehicles have a preference for the latter. Combining the two alternatives (using resistor plugs, resistance wire, and external resistors) is not recommended.

BYPASS CAPACITORS IN THE IGNITION SYSTEM

To reduce the conducted noise in the ignition system, bypass capacitors can be used. One 0.5-mfd bypass capacitor installed at the battery terminal of the ignition coil (Fig. 2-7) and another placed at the accessory terminal of the ignition switch have been found effective in a great many cases. In fact, many installations require only minor noise-suppression measures, such as a necessary resistor in the distributor lead and bypass capacitors at the generator and the regulator.

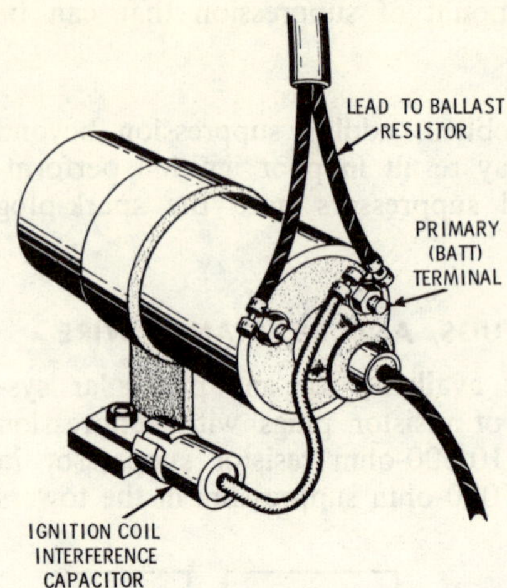

Fig. 2-7. Bypass capacitor on an ignition coil.

If further suppression is needed, it is best to determine how the noise is entering the receiver. This will allow a better evaluation of the best way to proceed. For example, if it is determined that the noise is entering via the power lead, adding a bypass at the accessory terminal of the ignition switch or bonding the instrument panel is more likely to produce results than using hood bonds. On the other hand, hood bonds would be a better approach if the noise is being radiated into the antenna.

Radiated noise may be present in the primary circuit of the ignition coil and in the electrical system of the auto. These noises are produced in the high-tension lead, and appear in the electrical system only as a secondary effect. However, there is a type of interference that is produced in the primary circuit of the ignition system and is coupled directly into the electrical system of the auto. This interference is low in frequency and is caused by the current pulses drawn by the primary of the ignition coil. The pulses appear on the power lead of the receiver because the wire may be common to it and the ignition coil.

Since this interference is low frequency, it is picked up in the audio section of the receiver rather than the r-f section or the antenna. It is easily identified, because the volume control has no effect on it. The treatment for this type of noise differs greatly from the normal ignition noise, so it is wise to check before proceeding too far with suppression measures. This type of noise can indicate a defective receiver. In older equipment, particularly, an open filter capacitor will let the noise reach the audio system. A check for this may save the money that would be spent unprofitably on suppression measures.

In new installations or in situations where the receiver is known to be in good condition, a 1000-mfd electrolytic capacitor can be installed at the ignition coil to cure the problem (Fig. 2-8). Since

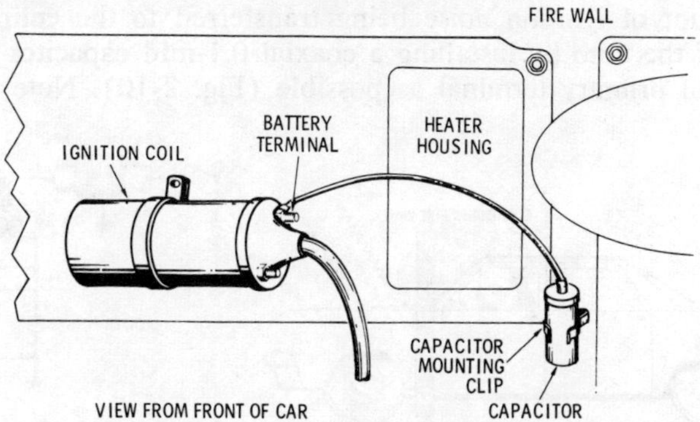

Fig. 2-8. Installating an electrolytic capacitor.

such a capacitor is expensive, it is good practice to use it only when the volume-control check indicates the problem is due to low-frequency noise. An alternate cure might be to run the receiver power lead directly to the battery terminal.

There are two schools of thought on connecting mobile communications equipment of any kind to power. One school says to run the power lead directly to the battery, thus minimizing the effect of conducted noise from the ignition system. The other school says that connecting directly to the battery may cause trouble since the power lead would then be exposed to the radiated noises present in the engine compartment. The author recommends that you try both methods as part of any noise-suppression program. Use the method which gives the least noise in your particular situation.

In many installations it is effective to install a bypass capacitor and/or a filter at the point where the power lead is connected. In

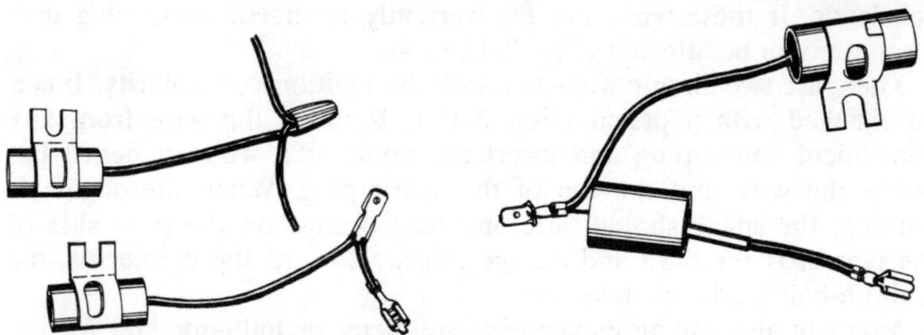

Fig. 2-9. Power leads with connectors and capacitors.

that case commercial assemblies that include the necessary power lead and connectors are often convenient to use. Typical examples of such assemblies are shown in Fig. 2-9.

In cases of severe interference it is possible to greatly reduce the amount of ignition noise being transferred to the entire wiring system of the auto by installing a coaxial 0.1-mfd capacitor as close to the coil primary terminal as possible (Fig. 2-10). Note that this

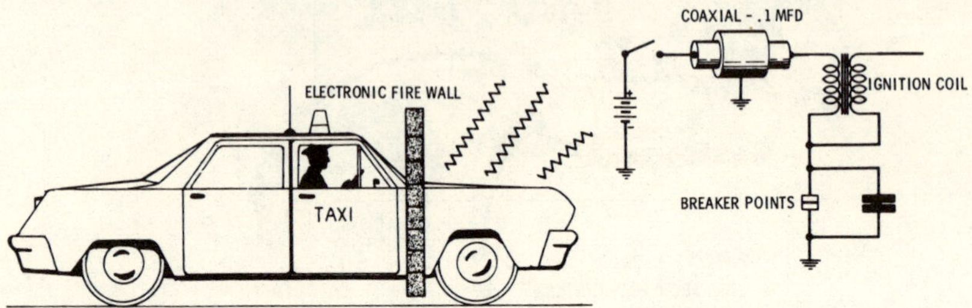

Fig. 2-10. Installing a feedthrough filter.

is not the coil-distributor terminal. Do not use a conventional bypass capacitor; this calls for a filter-type feedthrough coaxial. Ordinary bypass capacitors are unsatisfactory in this application since they are effective within an extremely limited frequency range (3 to 5 mc) when installed in the primary circuit.

When installing a feedthrough capacitor, remove the ignition coil and its mounting bracket also. Clean the paint off of the coil mounting bracket and the engine block, and then reassemble tightly. In many cases this will greatly reduce radiated interference.

CORRECT COIL POLARITY

Spark plugs require less voltage when the ignition-coil polarity is correct. When changes are made to the ignition circuit (addition of suppression, shielding, etc.), it may be necessary to disconnect the coil leads. If these wires are inadvertently reversed, spark-plug performance will be affected (Fig. 2-11).

There are two simple ways to check the ignition-coil polarity. It can be checked with a pencil (Fig. 2-12). Remove the wire from any convenient spark plug and insert the point of a wooden pencil between the wire and the top of the spark plug. When the engine is running, the spark should flare and turn orange on the plug side of the pencil. If the flare and orange tint are toward the connector, the ignition-coil leads are reversed.

You can also use an inexpensive voltmeter or milliammeter for the polarity check. Connect the positive meter terminal to the engine

26

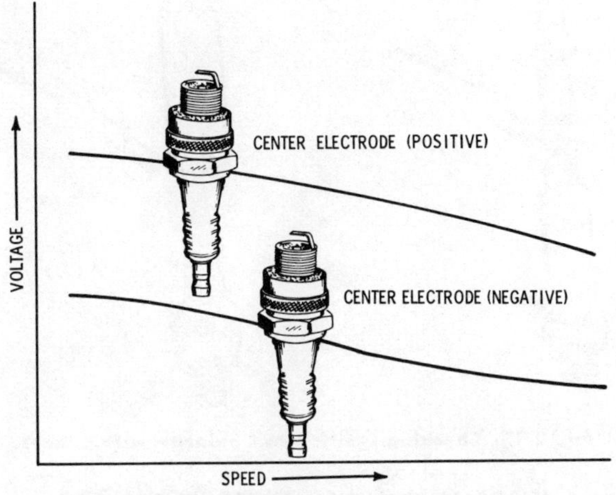

Fig. 2-11. Spark-plug firing characteristics.

ground and fasten a resistor of approximately 100,000 ohms to the negative test prod (Fig. 2-13). Start the engine and touch the end of the resistor momentarily to the spark-plug connector. If the needle kicks up scale, the polarity is correct. If the meter moves down scale, the coil leads are reversed. (The spark plug will misfire momentarily when contact is made.)

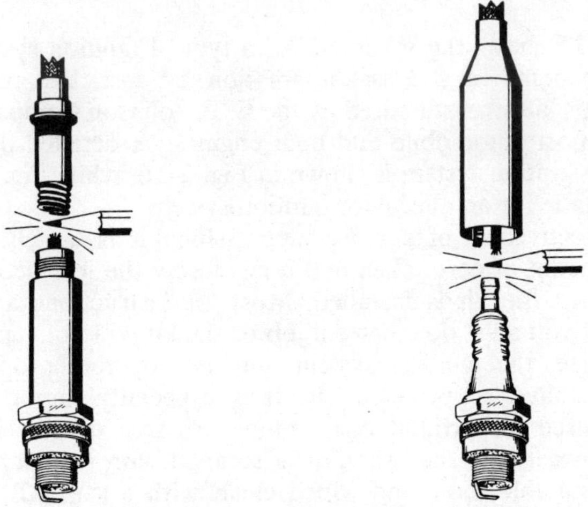

Fig. 2-12. Checking ignition-coil polarity with a pencil.

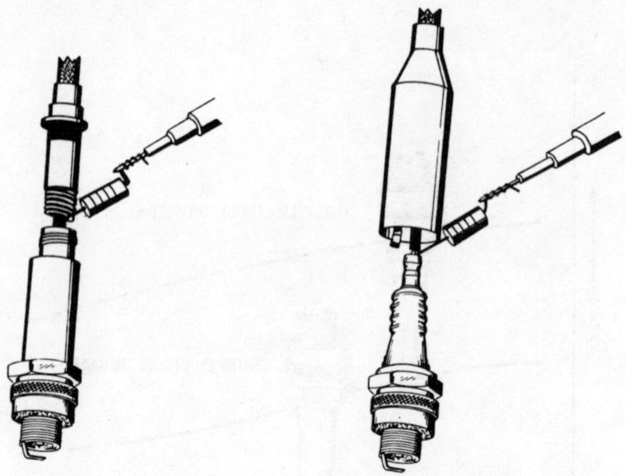

Fig. 2-13. Checking ignition-coil polarity with a meter.

INSTALLING NOISE-SUPPRESSION KITS

There are noise-suppression kits available that provide all of the components necessary for the usual suppression techniques. They supply suppressors for each of the spark plugs, a suppressor for the distributor, and filter capacitors for various points throughout the ignition system. Shielding for generator or alternator wiring is also included (Fig. 2-14). The following paragraphs describe the steps in the installation of a typical noise-suppression kit. In studying these steps the reader should become familiar with sources of radiated and conducted ignition noise, as well as how to suppress or filter these sources.

Fig. 2-15 shows the schematic of a typical ignition system with all the components of a noise-suppression kit installed at the proper points. This kit, manufactured by the E. F. Johnson Company, is available for most automobile and boat engines. A detailed diagram of a generator ignition system is shown in Fig. 2-16, while Fig. 2-17 shows the schematic for an alternator ignition system.

The effectiveness of a noise-suppression kit is largely determined by the amount of care taken and how closely the instructions are followed when the kit is installed. Most kit instructions are well prepared, but you must do a careful job or the kit will not suppress noise. Even worse, the ignition system may not operate properly with a poorly installed suppression kit. It is especially important that all surfaces used for ground connections are free of grease, dirt, and paint. If necessary, they should be scraped with sandpaper, a knife, or other suitable tools and wiped clean with a rag. All ends of the shielded braid (tails) in the kit have been kept as short as possible.

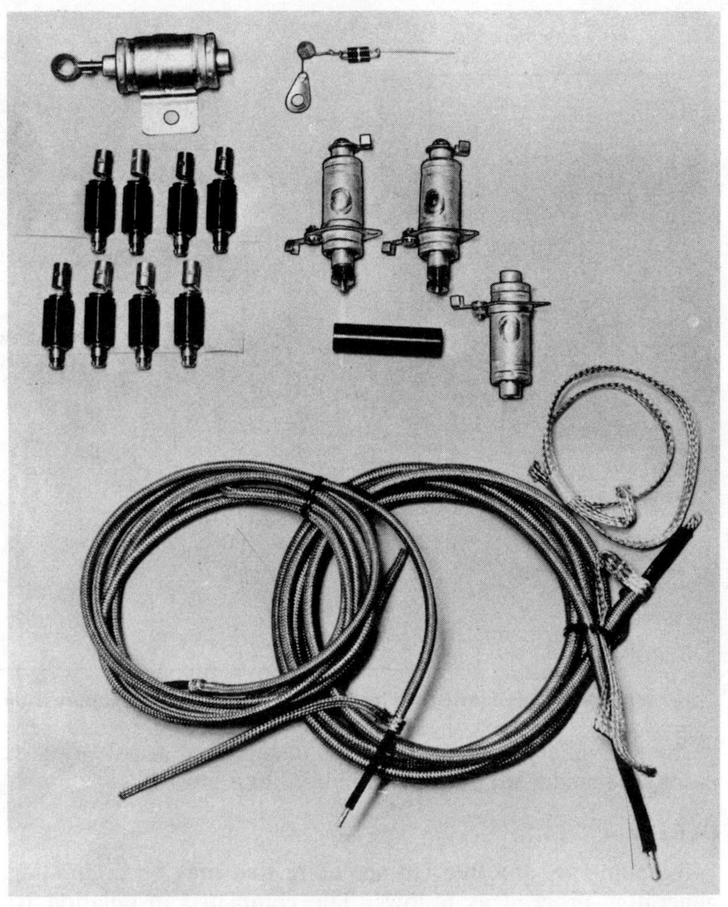

Fig. 2-14. Noise-suppression kit.

All connections to ground should be tightened securely. Care should be taken to insure that all wires removed from terminals are reconnected to their proper terminals. It is best to work with only one wire (or the set of wires from one terminal) at a time to prevent confusion. One of the battery cables should be removed to prevent any damage that might result from the accidental short circuits. The following items should be located before starting the installation: ignition coil, distributor, voltage regulator, generator (alternator), and battery. All hardware and solder lugs supplied with the kit should be left in their marked packages until they are used so that they can be properly identified. Refer frequently to the illustrations provided

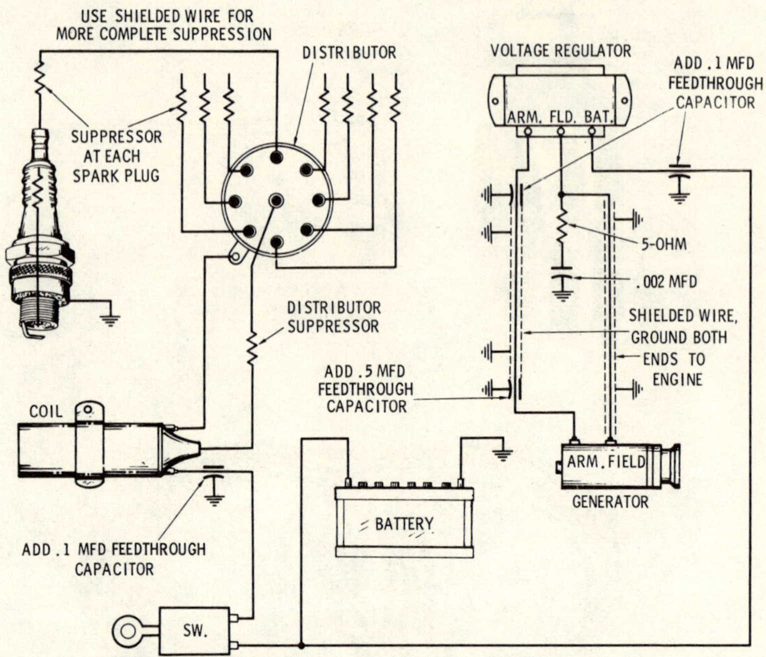

Fig. 2-15. Schematic for electrical system with noise suppression.

in the kit while you are making the installation. A soldering iron and rosin-core solder are required for installing most kits.

Generator

To eliminate any interference noise that may be originating in the generator, proceed as follows. The completed installation is shown in Fig. 2-18.

1. Disconnect one of the battery cables to prevent short circuiting the battery while the noise suppression kit is being installed.
2. Remove any existing radio noise suppression capacitors from the armature terminal (usually marked "A" or "ARM") of the generator. Usually these are small round metal cans with a single insulated wire (Fig. 2-19). If there are any of these capacitors on the voltage regulator or ignition coil, they also should be removed. Coaxial capacitors to be installed eliminate the need for them.
3. Mount an 0.5-mfd coaxial capacitor (Fig. 2-18) on the generator case with the screw that held the old capacitor or with one of the case bolts. Transfer the wire on the armature terminal to the furthest terminal of the coaxial capacitor.

30

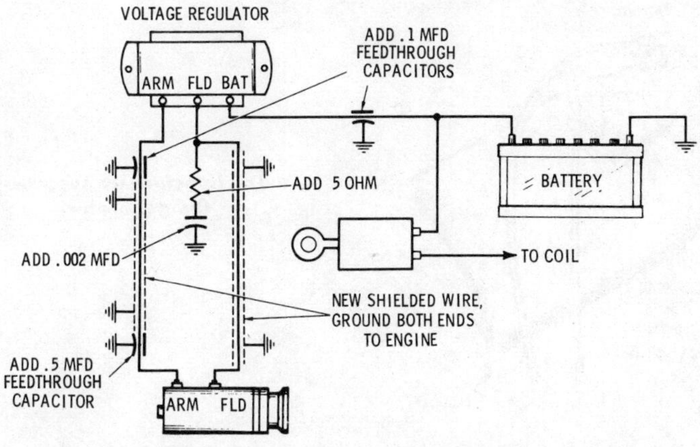

Fig. 2-16. Schematic for ignition circuit using a generator.

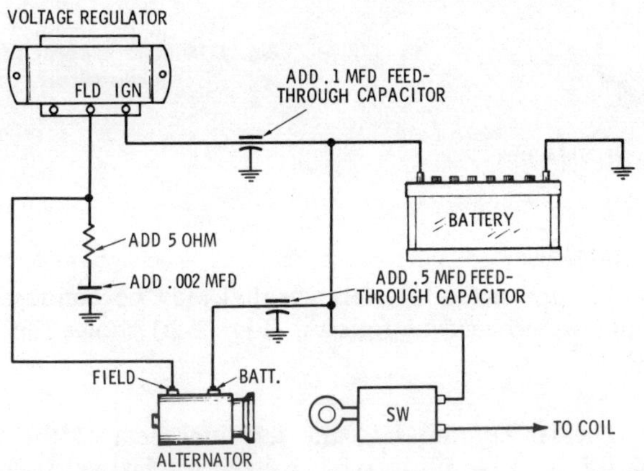

Fig. 2-17. Schematic for ignition circuit using an alternator.

4. If the generator permits the coaxial capacitor to be mounted close to the armature terminal, thread a 10-32 stud into the unused capacitor terminal and tighten it securely. Wrap the ears of a 1/4-inch solder lug around the stud and attach it to the armature terminal. Solder the lug to the stud. Part of the stud should be cut off if it is too long.

5. If the coaxial capacitor cannot be connected to the armature terminal of the generator with a stud and a solder lug, make a lead from the shortest possible length of shield braid. A solder lug should be crimped and soldered to each end of the braid. Use the shield braid supplied with the kit.

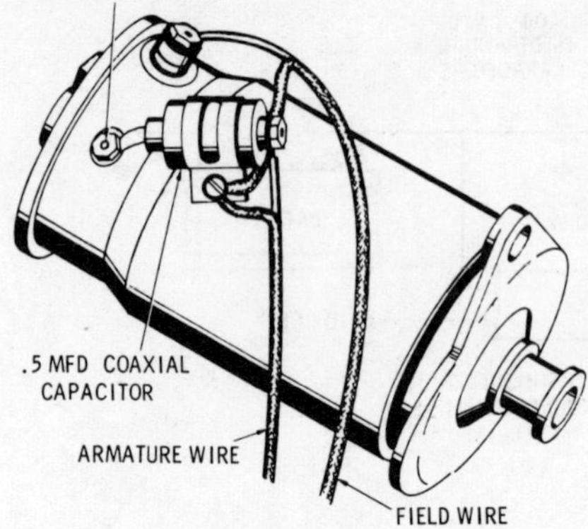

Fig. 2-18. Interference suppression at the generator.

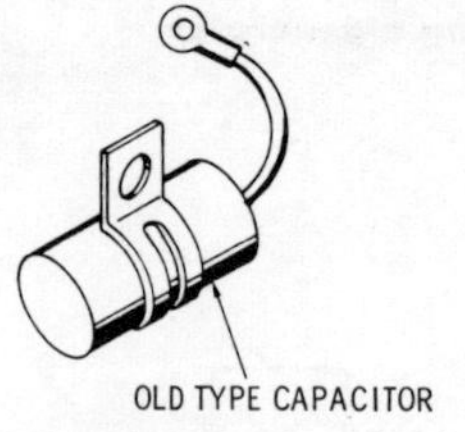

Fig. 2-19. Old type of suppressor capacitor.

Voltage Regulator

To eliminate any noise interference that may be coming from the voltage regulator, proceed as follows. (Fig. 2-20 shows the complete installation.)

1. Thread a 10-32 stud into the terminal nearest the mounting flange of a 0.1-mfd coaxial capacitor; fasten a shakeproof washer against the capacitor with a 10-32 nut. Then thread another 10-32 nut onto the stud finger-tight.

2. Remove the wires from the armature (usually marked "ARM") of the regulator. Either remove the regulator cover or bend the two mounting flanges of the coaxial capacitor so they will clear the regulator cover. (If the cover is removed, be sure to replace it later exactly as it came off.)

3. Place a number 10 shakeproof washer on the capacitor stud, and thread the stud into the armature terminal of the regulator (Fig. 2-20). When the stud protrudes from the other side of the terminal, tighten the nut and washer against the terminal. Do not let the stud touch the regulator mounting flange (ground).

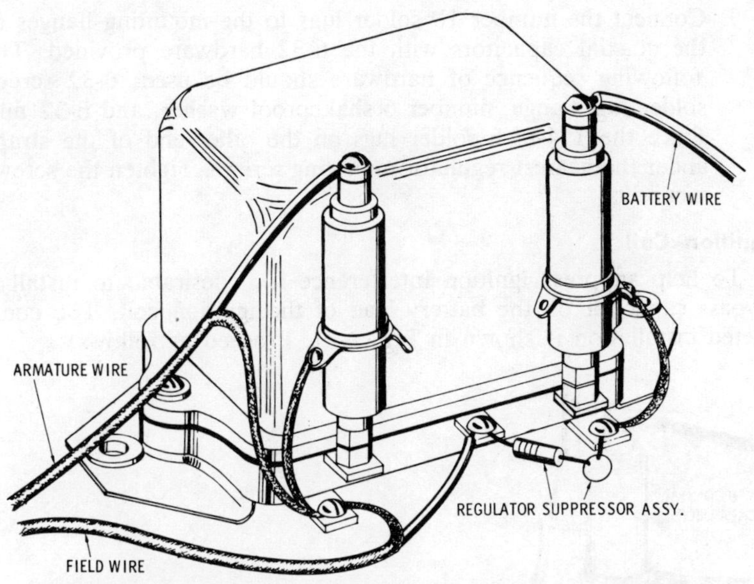

Fig. 2-20. Interference suppression at the voltage regulator.

4. Connect the wires removed from the armature terminal of the regulator to the unused terminal of the coaxial capacitor just mounted.

5. Remove the wires from the battery terminal (usually marked "BAT") of the regulator, and install a 0.1-mfd coaxial capacitor on the battery terminal in the same manner as steps 2 and 3.

6. Connect the wires removed from the battery terminal of the regulator to the unused terminal of the coaxial capacitor just mounted.

7. Place the solder lug of the regulator suppressor assembly (the 5-ohm resistor and 0.002-mfd capacitor) under the regulator mounting screw nearest the field terminal (usually marked "FLD") of the regulator. Connect the wire to the field terminal of the regulator by wrapping it around the screw. Keep the wire as short as possible and cut off any excess.

8. Make up two ground straps for the coaxial capacitors mounted on the regulator from shield braid. To do this crimp and solder the ears of a number 10 solder lug onto the braid. Determine the length of shield braid necessary to reach from one of the mounting holes in the flange of the capacitor to the nearest regulator mounting screw. Use the nearest hole in the capacitor flange to keep the lead as short as possible. Solder a 1/4-inch lug with ears to the other end of the braid.

9. Connect the number 10 solder lugs to the mounting flanges of the coaxial capacitors with the 6-32 hardware provided. The following sequence of hardware should be used: 6-32 screw, solder lug, flange, number 6 shakeproof washer, and 6-32 nut. Place the 1/4-inch solder lugs on the other end of the straps under the nearest regulator mounting screws. Tighten the screws securely.

Ignition Coil

To help suppress ignition interference it is desirable to install a bypass capacitor on the battery lead of the ignition coil. The completed installation is shown in Fig. 2-21. Proceed as follows:

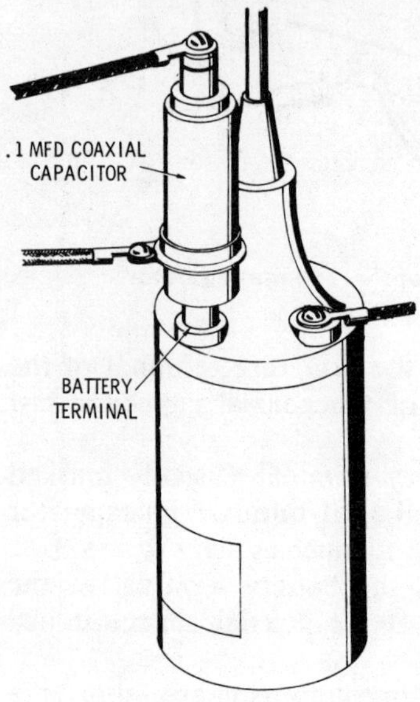

Fig. 2-21. Interference suppression at the ignition coil.

1. Remove the wire from the battery terminal (usually marked "BAT") of the coil. Place a number 10 shakeproof washer on the terminal and thread a 0.1-mfd coaxial capacitor onto the terminal. It will probably be necessary to bend the mounting flanges of the capacitor so that it can be threaded onto the terminal. Invert the capacitor if necessary. Tighten securely.
2. Connect the wire from the battery terminal to the unused terminal of the capacitor just mounted.
3. Crimp a number 10 solder lug with ears onto one end of some shield braid, and solder it.
4. Use the number 6 hardware provided to attach the solder lug to the mounting flange on the coaxial capacitor just mounted. Use the mounting-flange hole nearest the engine.

5. Run the shield braid to the nearest bolt on the engine that can be used for a ground connection. Fasten a solder lug of suitable size to the other end of the shield braid. Solder the connection and tighten the bolt securely. Keep this shield-braid strap as short as possible.

Shielded Generator Wires

Wires connected to armature and field terminals of the generator should be shielded to prevent the radiation of commutator hash. It is usually not necessary to shield these wires on engines using alternators. The shielded wires in the kit are 8 feet long to accommodate the longer runs required by some engines. However, if less wire is required between the generator and regulator, the shielded wires should be cut. To determine the length of shielded wires required, it will probably be necessary to lay the wires along the intended route. If both the generator and the regulator are on the same side of the engine, the routing of the wires will probably be quite short and simple. If they are on opposite sides of the engine, it will be necessary to route the wires around the engine, usually along the firewall, above the radiator, or below it. It is best to follow the regular cabling whenever possible since there are usually cable clamps which may be used to hold the new shielded wires. Be sure to allow a little extra for the slack needed at each end to make proper connections to the terminals. The new ends of the shielded wires should be prepared the same as the ends that were cut off. The shield-braid tails (ground straps) should be unsoldered from the unused portion of the wires and resoldered to the new ends of the shielded wires. Be sure to tape the shield braid to the wires at both ends with either electrical tape or friction tape. This will prevent the shield braid from sliding on the wire and touching the bare terminals, thus causing a short. When the leads are the proper length, proceed as follows:

1. Strip off 3/8-inch of insulation and place a number 10 flag lug on each end of the shielded armature wire. This is the larger of the two shielded wires; it runs between the coaxial capacitor on the armature terminal of the generator and the coaxial capacitor on the armature terminal of the voltage regulator. Crimp the flags of the solder lugs tightly around the bare wires and solder.
2. Check the size of the generator field terminals and fasten a solder lug of suitable size to the generator end of the shielded field wire. This is the smaller of the two shielded wires (usually number 14 wire); it runs between the field terminal of the regulator and the field terminal of the generator. Strip 3/8-inch of

insulation off the other end, crimp a solder lug onto the bare wire, and solder it.

3. Cut the shield-braid tails to their proper lengths. This can be done more accurately if the center conductors are first connected to their proper terminals—the large wire to the armature terminals and the smaller to the field terminals. At the regulator the shield braids are connected with 1/4-inch lugs to the nearest mounting screws of the regulator (Fig. 2-20). At the generator the shield-braid tails are connected, with 1/4-inch solder lugs, either under the screw mounting the coaxial capacitor on the generator or under the nearest generator case bolt. While the shield-braid tails should be kept as short as possible, be sure they clear the terminals on the generator and regulator (Figs. 2-18 and 2-20). Make sure the ground points are free of paint and dirt. The shield-braid tails may be covered with tape or plastic tubing if necessary.

4. It is advisable to either tape or tie the shielded wires down at various point along their route, especially if they pass near any moving parts. Use existing cable clamps wherever possible. Reconnect the battery cable that was removed from the battery earlier. Be sure it is tightened securely. The original wires which are no longer used should be secured by tape or other means if it is not practical to remove them.

Distributor Suppressor

Adding a suppressor at the distributor is a simple task and can be quickly done. Cut the distributor rotor wire in two about 3/4 inch above the distributor cap. Insert the suppressor by twisting the cut ends of the wire into the ends of the suppressor (Fig. 2-2). When the high-voltage wire is returned to the center terminal of the distributor cap, make sure it is pushed all the way down.

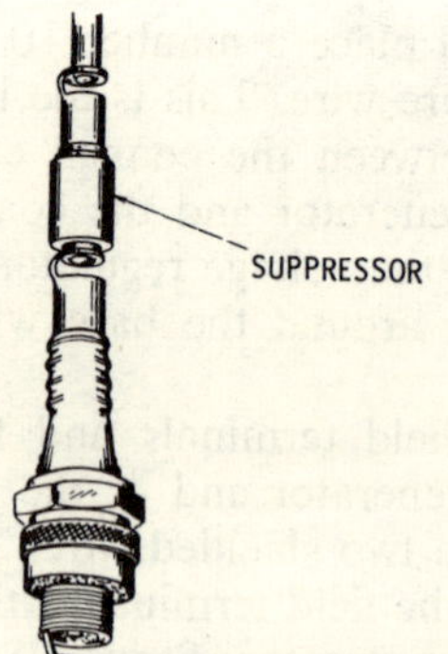

Fig. 2-22. Spark-plug suppressor installation.

Spark-Plug Suppressors

Install the spark-plug suppressors on each of the spark plugs, removing only one spark-plug wire at a time. Connect the spark-plug wires to the other end of the suppressors (Fig. 2-22). Be sure all the connections are snug fitting to prevent external arcing. Do not use spark-plug suppressors if resistor spark plugs have been installed in the engine.

CHAPTER 3

Suppressing Other Engine Interference

There are many sources of interference besides the basic ignition circuit. Both the generator and voltage regulator can contribute to noise problems, and each must be handled in a special way. All possible sources of noise are not eliminated even in the latest transistor ignition systems. Maintenance of spark plugs, the distributor, and the ignition coil plays an important part in the elimination of interference, so each of these components must be examined. It is also necessary to study procedures and equipment for locating interference at any point in the engine system.

LOCATING INTERFERENCE SOURCES

There are many ways in which noise or interference can be traced back to its source. Some of these are fairly elaborate, while others are quite simple. If an oscilloscope is available for the noise-tracing job, connect a pair of long, shielded leads to the vertical input of the scope, and attach a pickup coil (for radiated noise) or a probe (for conducted noise) to the shielded leads. The scope can be connected through an isolating capacitor to possible points of conducted noise (power leads into the receiver, generator output, ignition switch hot lead, instrument hot leads, etc.) instead of using a standard probe. A signal tracer can also be used to check various points for conducted noise.

If neither one of these instruments is available, a receiver itself will usually serve as an excellent noise tracer, particularly for radiated noise. The receiver can be converted into a noise tracer by connecting a pickup coil to the antenna input terminals. A suitable pickup coil consists of 50 turns of insulated wire wound into a 2-inch coil and taped to the end of a broom handle (Fig. 3-1). The coil is connected to the receiver antenna through ordinary lamp cord or any

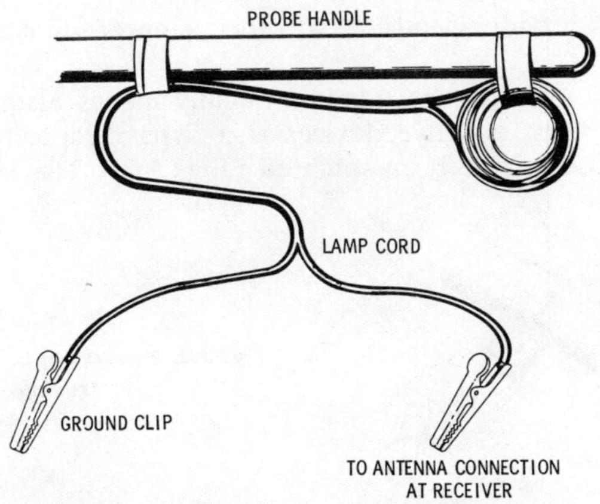

Fig. 3-1. Pickup coil for noise tracing.

suitable insulated wire. Disconnect the regular antenna. If the receiver has a coaxial input, it will be necessary to make up a coax adapter in order to connect the coil.

No matter what instrument you use, the basic procedure for locating radiated noise is as follows:

1. Operate the engine and hold the pickup coil near the distributor. Listen or watch for a change in the noise level.
2. Move the pickup coil along each spark-plug lead from the distributor to the individual spark plugs.
3. Move the pickup coil along the high-tension lead from the distributor to the ignition coil.

If a measurable amount of noise is detected anywhere near the ignition system, it is possible that either a partial or a complete shielding system will be required to eliminate the noise.

In many cases a dip meter can be used to trace a source of radiated noise. Usually radiated noise covers a wide range of frequencies; if the noise falls in the range of a dip meter, it can be detected and identified. The dip meter should be operated as a diode detector with the headphones plugged in. On most dip meters this means setting the oscillator-diode switch to diode and plugging the headphones into the appropriate jack on the meter. Hold the dip-meter pickup coil near the suspected point of radiated noise (distributor, spark-plugs, spark-plug leads, ignition coil, etc.). Slowly tune the dip meter over its entire operating range while listening for noise in the headphones. If the noise appears only on one frequency, it is possible that a wave trap can be used. However, radiated noise usually occurs across a wide spectrum, so other suppression techniques are required.

Conducted noise can be traced by many means also. One of the simplest and most effective devices is a bypass capacitor, to which two alligator clips have been soldered (Fig. 3-2). The large alligator

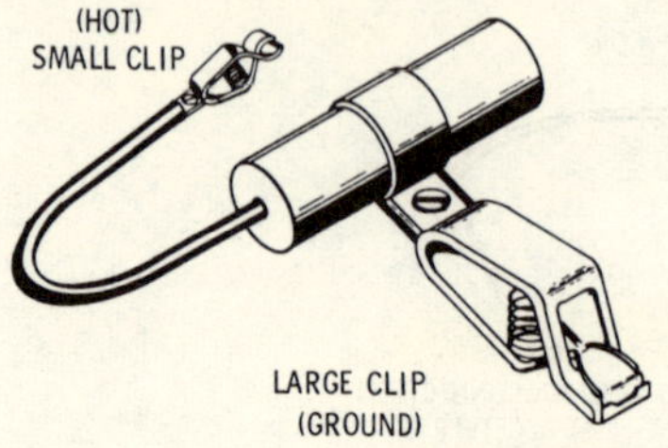

Fig. 3-2. Bypass capacitor for noise tracing.

clip is attached to the capacitor shell, and the smaller clip is attached to the lead. To use such a capacitor in noise tracing, connect the small clip to the circuit at a point near a suspected source of noise. Then connect the large clip to the nearest ground point. If there is any noticeable reduction in noise level with the bypass capacitor connected, install a permanent capacitor at that point. Use the same value for the permanent capacitor as is used for the test bypass. It should be between 0.5 and 1 mfd for best results.

THE GENERATOR

Generator noise is caused by current arcing between the generator brushes and the armature. The noise may be identified as a high-pitched whine that varies with the speed of the engine. Positive identification can be made by disengaging the generator from the fan belt. If the noise is gone when the engine is running without the generator, it may be assumed that the generator is the source of the noise.

Generator noise is usually eliminated when an 0.5-mfd bypass capacitor is installed at the armature terminal (Fig. 3-3). If the noise is severe, a 1-mfd capacitor may be required. In an older automobile it may be necessary to replace worn brushes and possibly to dress the commutator. It is usually easier to replace generator brushes when they are causing noise than it is to bypass the noise by adding a capacitor. However, do not attempt any extensive generator repair work unless you are sure that the generator is the noise source, and you are qualified to do generator work. You can end up with worse troubles than noise if you should damage the generator during repair. Follow the manufacturer's recommendations on all generator repairs. If in doubt, have the generator checked and repaired by a qualified mechanic or auto electrician.

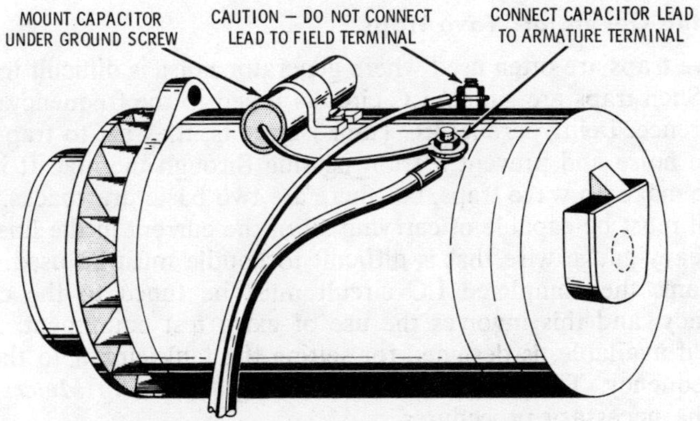

Fig. 3-3. Bypass capacitor on generator.

WARNING

Never connect a bypass capacitor to the field lead of a generator because this can damage the generator beyond repair. The armature terminal to which the capacitor is to be attached can usually be identified by the heavy lead connected to it. Most domestic autos have a warning label at the field terminal. When in doubt, consult the auto service manual.

In the case of excessive generator noise, it may be of some help to use a coaxial capacitor instead of the bypass (Fig. 3-4). A similar coaxial capacitor can also be used at the voltage-regulator terminals,

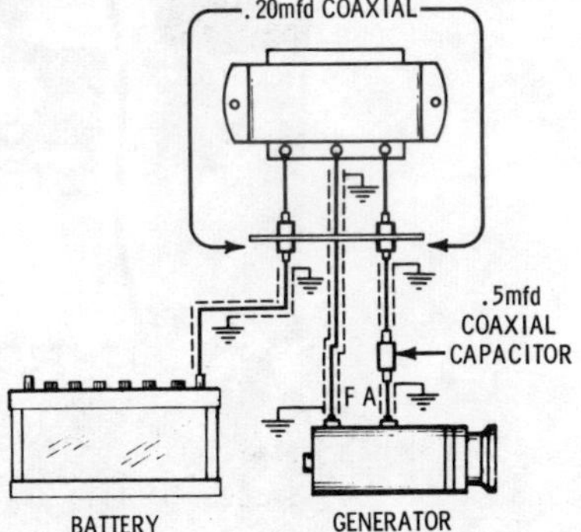

Fig. 3-4. Coaxial bypass-capacitor installation.

or a dual unit can be installed. Any factory-supplied bypass capacitor must be removed, but its replacement can be mounted at the same place. Run a short 10-gauge wire from the armature post of the generator to the coaxial capacitor, and connect the original wire to the other end of the capacitor.

Installing Generator Wave Traps

Wave traps are often used where generator noise is difficult to eliminate. Such traps are simple LC circuits tuned to the frequency of the interference. Being parallel LC (tank) circuits, they act to trap or reject the noise and prevent it from passing through the line. It is possible to make up wave traps, but there are two basic drawbacks. First, the coil must be capable of carrying all of the current in the line, so a very heavy-gauge wire that is difficult to handle must be used. More important, the completed LC circuit must be tuned to the correct frequency, and this involves the use of extra test equipment. A dip meter, if available, is designed for setting the tank circuit to the correct frequency. The author's book, *Servicing With Dip Meters,* contains the necessary procedures.

It is also possible to use commercial wave traps such as are shown in Fig. 3-5. These are manufactured by the Ben N. Bartlett Company in two sizes; the R/HD which carries 50 amperes, and the VR/C which is designed for smaller currents. The VR/C is normally installed at the field terminal of the voltage regulator and at the switch side of the ignition coil. The R/HD is used at the generator armature terminal and the voltage-regulator battery terminal. Some installations also require an R/HD filter at the armature terminal.

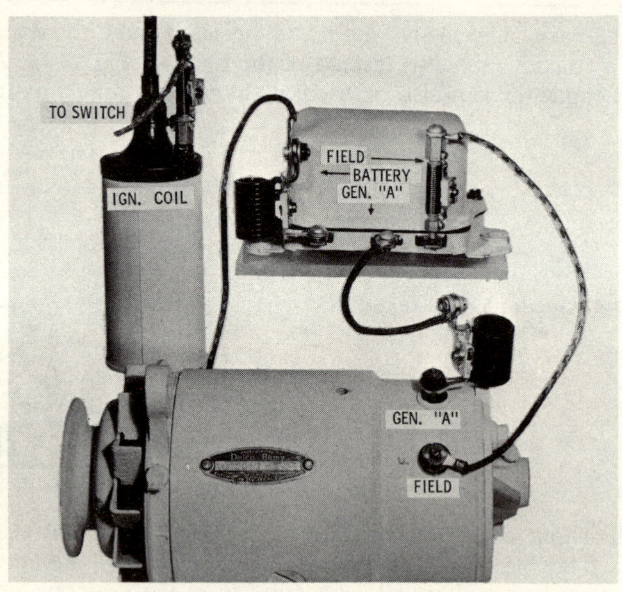

Courtesy Ben N. Bartlett Co.

Fig. 3-5 Wave-trap installations.

The filters shown in Fig. 3-5 are pretuned to the operating frequency of the communications equipment, usually at the approximate center of the band. Stock filters are available for the 11-meter Citizens band as well as the 2- and 6-meter amateur bands. Other filters can be obtained to cover special frequencies such as vhf-marine, industrial, and commercial bands—any frequency from 15 to 180 mc. The factory pretuned filters should not be adjusted after they are installed. The tuning adjustment (a trimmer capacitor) is set and sealed during manufacture.

THE ALTERNATOR

Alternators produce far less noise than generators, because alternators do not contain commutator segments. Instead, the brushes ride on continuous bands of copper (slip rings). The noise produced by alternators is usually a high-pitched whine similar to that of generators. You can verify the sourse of the noise by disengaging the fan belt. If the noise is gone when the engine is started, you are safe in assuming that the alternator is at fault. The diodes used to rectify the alternator output (to the direct current required by the battery and ignition system) are usually noiseless since they have no moving parts. On rare occasions, however, defective diodes can cause intermittent noise. Such diodes should be replaced. The voltage regulators are usually less of a noise problem than those for generators since the former contain only one relay, compared to three for the latter. Usually, a 0.1-mfd capacitor across the one set of regulator points suffices. If not, wave-trap filters can be installed as described for generators in the previous section.

THE VOLTAGE REGULATOR

Voltage regulators generally have terminals for connections to the generator armature, the generator field, and the battery. There are markings on the regulator cover to indicate which terminal is which.

The circuit breaker in the regulator is most likely to create radio interference. This interference can be identified as an irregular popping noise at idling speed. If the auto is equipped with an ammeter, the noise will coincide with a sharp movement of the ammeter needle. A 0.5-mfd capacitor, installed at the battery terminal of the regulator, will eliminate this particular noise (Fig. 3-6). This is normally the only suppression measure required at the voltage regulator.

Occasionally the relay which regulates the charging voltage may cause interference. The contacts of this relay are constantly opening and closing; arcing will produce a rasping noise in the receiver. In

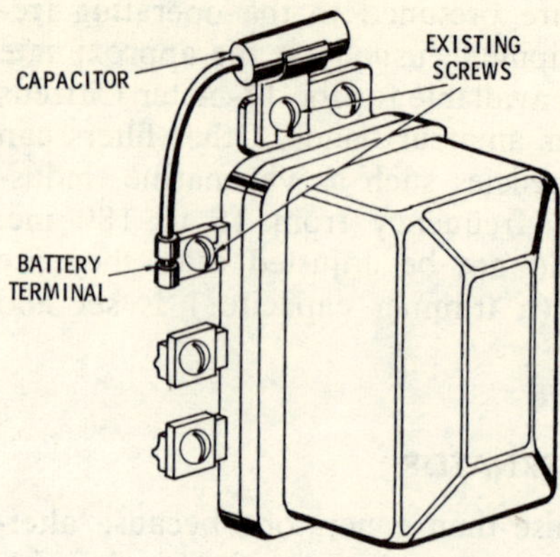

Fig. 3-6. Bypass capacitor on a voltage regulator.

this case, suppression is applied to the field terminal of the voltage regulator by a network (Fig. 3-7). The r-f choke is used only in severe cases; the capacitor and noninductive resistor normally are sufficient, but the resistor must be used with the capacitor. A bypass capacitor on this lead, at either the regulator or generator end, would damage the generator and could also damage the regulator.

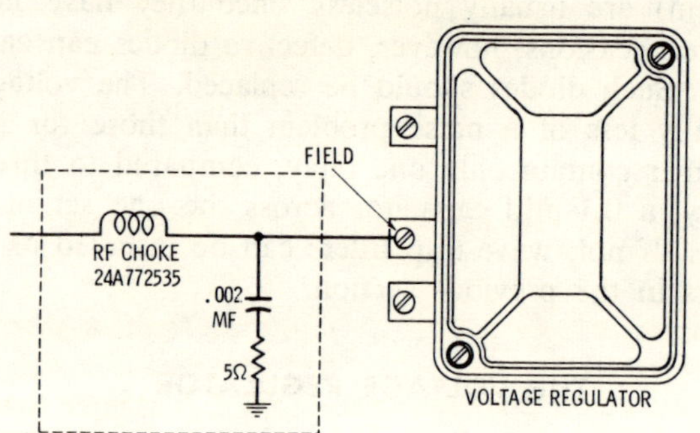

Fig. 3-7. Suppressor on a voltage regulator.

TRANSISTOR IGNITION SYSTEMS

The transistor ignition systems that frequently replace conventional systems also create interference since you still have a distributor, spark plugs, cables, and coils. All of these are capable of radiating noise of greater intensity due to the higher voltages available. The major interference problem develops when spark-plug gaps widen or the plugs deteriorate. The wider the gap, the higher the voltage is, and the more interference there is created.

As of this writing there are no kits or components designed specifically to eliminate noise in transistor ignition systems. However, there is one transistor ignition system with built-in noise suppression, filtering, and partial shielding. This unit is shown in Fig. 3-8. It is completely self contained—fully wired and shielded—and consists of a special high-voltage coil, a heavy-duty switching tran-

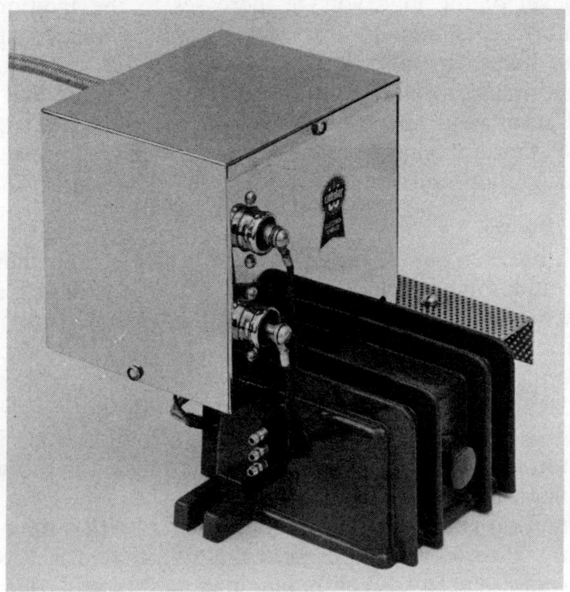

Fig. 3-8. Transistor ignition system with built-in shielding.

sistor mounted on a rugged heat sink, and the ballast-shorting relay. The casting also serves as a shielded housing for the circuitry. The high-tension coil is enclosed in a shielded container that prevents direct radiation of ignition-noise impulses. As further protection, the low-voltage leads from the coil pass through coaxial capacitors that effectively filter out noise pulses from the external circuits. The high-voltage leads from the coil are enclosed in a shielded braid.

An internal relay shorts out the ballast resistor present in all 12-volt ignition systems without a need for direct wiring to the ignition switch. This must be shorted out for transistor systems, an extremely difficult procedure in most modern vehicles. In addition, the unit uses a patch plug for simple connection to existing ignition-circuit primary leads. The original ignition system remains intact and can be put back into operation merely by shifting the patching connector. This feature provides a means for comparing performance and offers an extra safety factor. The complete assembly can be mounted readily on the firewall or other area in the engine compartment by flanges on the casting.

ELECTRICAL MAINTENANCE

As the level of the high voltage increases in the ignition system, stronger interference signals are radiated. Worn, ragged spark-plug gaps require higher ignition voltage, so both radio performance and engine performance deteriorate. Flat, parallel gaps and factory-maintained settings in spark plugs require less voltage and decrease interference. Careful spark-plug maintenance is therefore the first step toward ignition-noise suppression.

Distributor Points

Although breaker points emit a low-frequency signal, they are not generally troublesome beyond the standard-broadcast band. However, point bounce is detrimental to high-speed engine performance, and the intermittent fluttering of the contacts can broadcast rapid, high-voltage, high-frequency interference over the entire radio spectrum. The secondary wiring acts as the radiating agent, with each different length of wire emitting a different frequency. Point bounce is usually caused by too much spring tension. If spring tension is too weak, the points will float, and this also is undesirable.

Point condition is important for ignition circuits that have suppression systems, since a fully suppressed ignition system is less tolerant of breaker-point neglect. Well-maintained points will decrease the possibility of noise signals and provide optimum engine performance.

You can often improve point performance by matching the capacitance value of the distributor condenser to your type of driving. This avoids premature pitting of the points. The distributor condensers for most autos have an allowable capacitance range of 0.18 mfd to 0.2 mfd. If you do a lot of stop-start driving, use a value near the high limit. If most of your driving is at high speed on the freeway, use a low value. As illustrated in Fig. 3-9, too low a capacity generally shows up as a crater in the negative contact, and too high a capacity results in a crater in the positive contact.

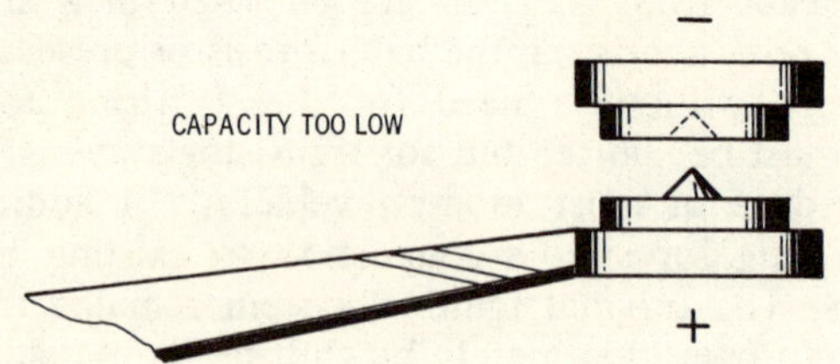

Fig. 3-9. Effects of improper distributor-capacitor values.

Before you adjust the condenser capacities to fit your type of operation, look for loose primary connections or abnormal generator charging rates. Malfunctions in the ignition system can also cause pitted contacts.

Distributor Cap and Rotor

Rotors and distributor caps should be replaced every 25,000 miles. Never file a rotor tip; replace the rotor. Also, remember that rotor gaps cannot be measured accurately. After 25,000 miles the inside of the distributor cap is impregnated with millions of metallic particles thrown off by the rotor. These act to radiate noise. To keep interference at a minimum, replace the cap whenever the rotor is replaced.

Ignition Coil

The high-tension terminal socket of the ignition coil may be corroded as a result of arcing caused by failure to insert the end of the cable into the socket properly. This additional arcing is one more source of interference. Corrosion may also develop in coastal areas due to salt air. Since any corrosion will cause resistance to current, the socket should be thoroughly cleaned out with a terminal cleaner, sandpaper, or a stiff wire brush. The cable terminal should be cleaned with sandpaper.

Spark Plugs

Carefully inspect the insulators and electrodes of all spark plugs. Replace any plug that has a cracked insulator, a broken insulator, or a loose electrode. If the insulator is worn away around the center electrode, or if the electrodes are burned or worn so they cannot be adjusted for proper gap, the plug is worn out and should be discarded. When spark plugs are adjusted, use a wire feeler gauge of the

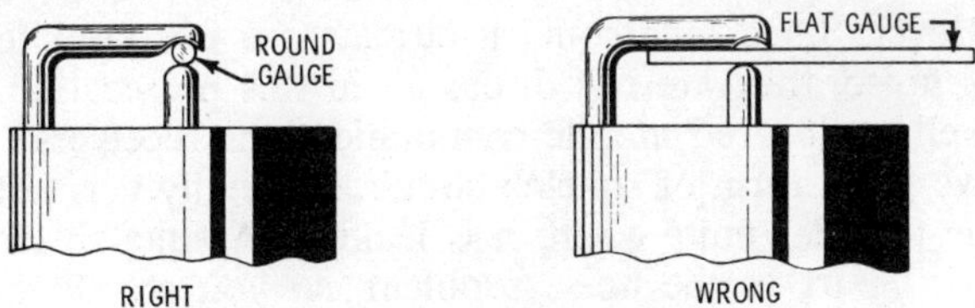

Fig. 3-10. Adjusting spark-plug gap.

diameter specified by the manufacturer. Flat feeler gauges do not give the correct measurement if the electrodes are worn (Fig. 3-10). Adjust the gap by bending the side electrode only; bending the center electrode will crack the insulator.

CHAPTER 4

Squelch Circuits

A squelch circuit is like a gate that will open on the arrival of a signal with certain predetermined characteristics and will close when that signal is removed. Usually the squelch remains closed until signals of sufficient strength are present. The squelch control sets the level of silence, or the signal level at which the receiver opens and closes. Many squelch circuits are provided with an override control or button that cuts out the squelch action so it is possible to listen for weak signals without changing the setting of the squelch control.

Practically all receivers used in mobile communications incorporate some type of squelch circuit. This applies both to auto and marine two-way communications systems. A squelch circuit will not eliminate noise (as does suppression, filtering, and shielding) or reduce the volume of noise (as does a noise limiter). However, it does provide complete silence until the desired signal is present.

The noise squelch circuits are not to be confused with tone squelch circuits. The latter keeps the receiver silent or off until a signal is received that is accompanied by an audio tone at a specific frequency. Noise squelch circuits are actuated by a change in noise or signal level. Since tone squelch circuits are not used to suppress or control noise, they are not discussed in this publication.

Most well engineered mobile-communications receivers of modern design have some form of squelch circuit. Normally it is not practical to add one to older units where it is lacking. A squelch circuit plays an important part in the noise problem, so you should be familiar with its operation.

SIGNAL SOURCES

There are three sources for signals to actuate squelch circuits—the negative automatic volume control (avc) voltage, the i-f amplifier screen voltage, and the background noise voltage.

The avc voltage is developed by the detector and is applied as a bias to the i-f stages. When no signal is being received, the negative avc bias voltage is low, and the i-f stages amplify a maximum amount. In other words, the receiver sensitivity is maximum. With a signal applied, the avc bias increases, reducing receiver sensitivity in proportion to the increase in signal strength. As a result the audio output of the receiver remains constant although the r-f signal received varies within a given range. The negative d-c voltage developed by the avc circuit can be used to actuate a noise-squelch circuit.

Instead of using the avc voltage directly, the screen voltage of an avc-controlled i-f stage can be used. When the negative avc voltage increases in the presence of a signal, the screen voltage of the i-f stage (or stages) also increases. (A negative voltage on a control grid decreases the screen current and allows the screen voltage to rise.) This positive d-c screen voltage can be used to actuate a noise-squelch circuit.

Noise picked up by the receiver can be detected and rectified into a d-c voltage. In the presence of a signal the background noise drops, as does the d-c voltage. These changes can be used to actuate a noise-squelch circuit.

SIGNAL POINTS

Logical points for obtaining a squelch control signal in a typical receiver are indicated in Fig. 4-1. The avc voltage can be picked off at points 1 or 9. The i-f amplifier screen (point 3) is the most logical

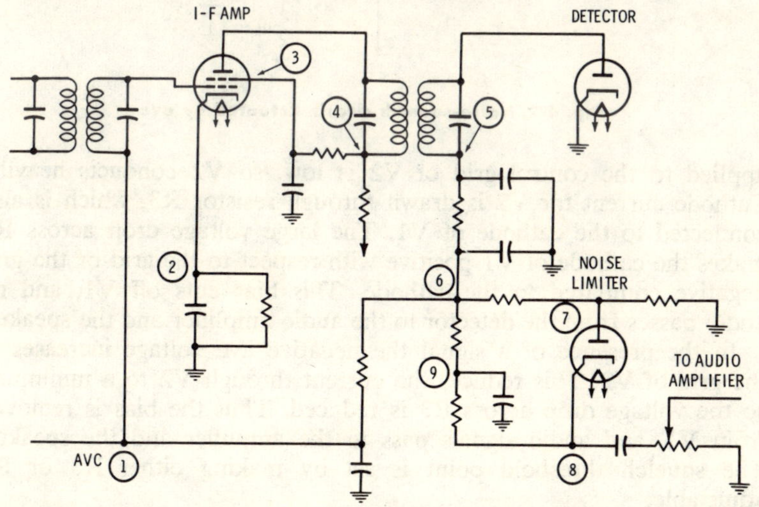

Fig. 4-1. Signal pickoff points for squelch circuits.

pickoff point for a positive d-c voltage that will increase in the presence of a signal. The i-f plate (point 4) could be used, but the variation is usually too small to be practical. The i-f amplifier cathode (point 2) produces a positive voltage that decreases in the presence of a signal. The background noise can be picked off at points 5, 6, 7, or 8; points 5 and 6 are the most logical choices.

AVC-ACTUATED CIRCUIT

A noise-squelch circuit controlled by avc voltage is shown in Fig. 4-2. Tube V1 is an audio amplifier, while V2 provides the necessary squelch action. In the absence of a signal, the negative avc voltage

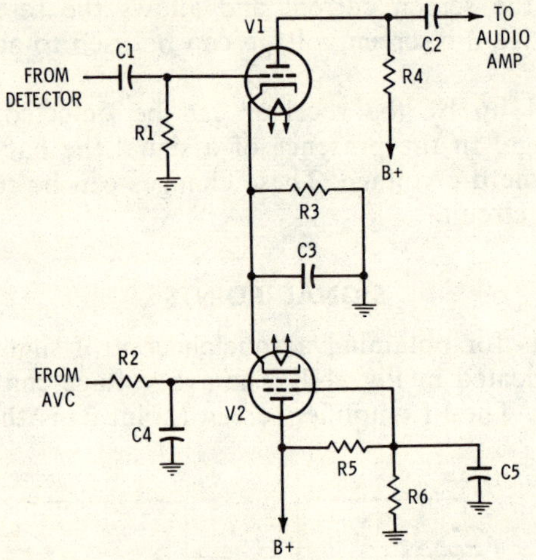

Fig. 4-2. Noise-squelch circuit actuated by avc.

applied to the control grid of V2 is low, so V2 conducts heavily. Cathode current for V2 is drawn through resistor R3, which is also connected to the cathode of V1. The large voltage drop across R3 makes the cathode of V1 positive with respect to the grid or the grid negative compared to the cathode. This bias cuts off V1, and no audio passes from the detector to the audio amplifier and the speaker.

In the presence of a signal the negative avc voltage increases at the grid of V2. This reduces the current through V2 to a minimum, so the voltage drop across R3 is reduced. Thus the bias is removed from V1, and audio signals pass to the amplifier and the speaker. The squelch threshold point is set by making either R3 or R5 adjustable.

Two noise-squelch circuits using the voltage from the screen of an i-f tube are shown in Figs. 4-3 and 4-4. Fig. 4-3 is a diode circuit, and Fig. 4-4 has a triode that controls the audio signal.

The diode of Fig. 4-3 can be either a tube or a crystal. In the absence of a signal, the cathode is more positive than the plate, so the diode does not conduct. The diode plate receives its voltage

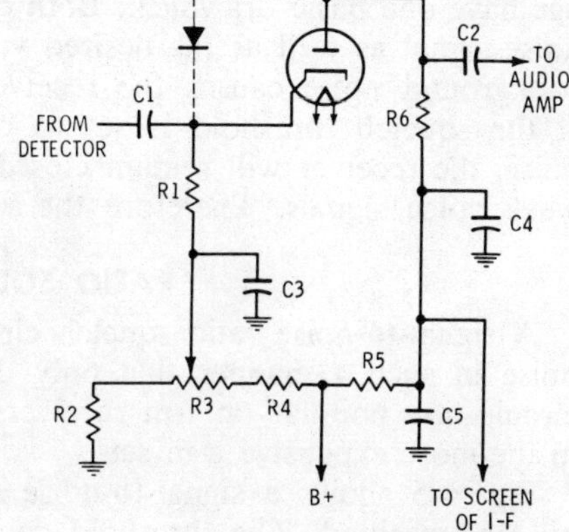

Fig. 4-3. Diode noise-squelch circuit.

through R6 from the screen of an i-f amplifier tube. Without a signal the screen draws a large current, causing a large voltage drop across R5. Therefore the i-f stage screen voltage and the diode plate voltage are low. The cathode of the diode receives its voltage through R1 from a divider consisting of resistors R2, R3, and R4. Resistor R3 is the squelch threshold control and permits the bias to be set as desired.

In the presence of a signal the screen voltage of the i-f amplifier (and the voltage at the junction of R5 and R6) rises. This increases the diode plate voltage above that of the cathode and permits the diode to conduct. Therefore, audio signals pass through to the audio amplifier stages and the speaker.

Triode V1 of Fig. 4-4 is biased so that the grid will be negative with respect to ground (and the cathode) in the absence of a signal. In this condition the triode does not conduct. The grid of V1 receives its voltage through R1 from a divider consisting of resistors R3, R4, R5, and R6. Resistor R3 is connected to a negative voltage and R6 is connected to B+. Resistor R4 is the squelch threshold control that permits the bias to be set as desired. The values of R3 through R6 are chosen so that, under no-signal conditions, the voltage on the slider of R4 is negative with respect to ground. The i-f screen is connected to the junction of R5 and R6.

In the presence of a signal the screen voltage rises, increasing the voltage at the junction of R5 and R6. This changes the drop across R3, R4, and R5 so that the bias voltage applied to V1 is now positive (or much less negative) in respect to ground. The change in grid bias permits V1 to conduct, passing audio from the detector to the audio-amplifier stages and the speaker.

Noise-squelch circuits that operate on avc voltage or i-f screen voltage have one basic drawback: Both circuits are actuated by a strong noise signal as well as the desired voice signal. A sharp increase in background noise causes the receiver to open and audio to pass. If the squelch threshold is set to overcome this high background noise, the receiver will remain closed to both background noise and weak voice signals. Therefore the weak signals will not be heard.

RATIO SQUELCH

A signal-to-noise ratio squelch circuit is operated by background noise in such a manner that only desired signals will pass. These circuits are popular on f-m receivers and are gradually finding use in the more expensive a-m sets.

Fig. 4-5 shows a signal-to-noise ratio squelch circuit developed by Hammarlund. The threshold control (R9) permits the operator

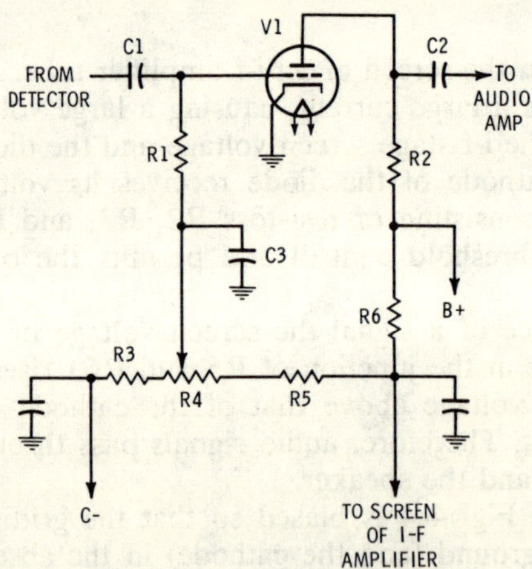

Fig. 4-4. Triode noise-squelch circuit.

to select the quality of signal in terms of signal-to-noise ratio that he wants to hear.

The detector output of the receiver is applied to a high-pass filter and a low-pass filter in order to separate the noise from the desired

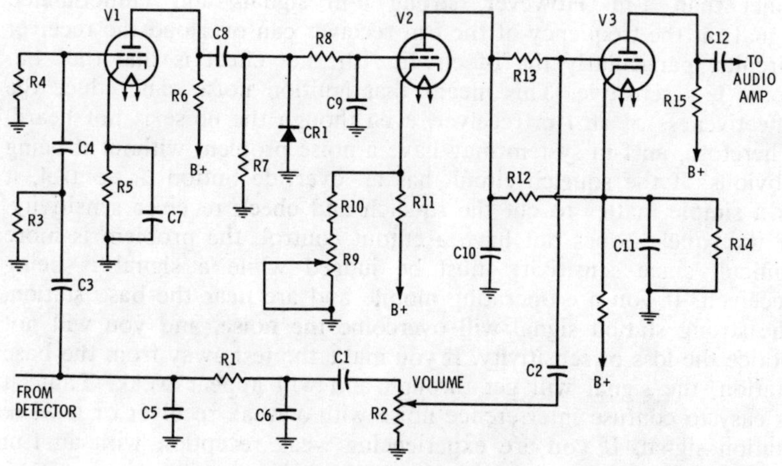

Fig. 4-5. Signal-to-noise ratio squelch circuit.

audio signals. Audio signals go through the low-pass filter (R1, C5, C6) and the volume control (R2) to the input of V3. When V3 is conducting, the desired audio signals pass through to the final audio-amplifier stages.

Background noise signals are applied through the high-pass filter (C3, C4, R3, R4) to the input of V1, an amplifier for the noise. The output of VI is applied to the input of V2. This input is always positive, because the negative noise peaks are shorted to ground by CR1. The amount V2 conducts is related to the noise received; the point at which V2 starts to conduct is determined by squelch thresh-hold control R9 that sets the V2 cathode voltage. As long as V2 is conducting, its plate voltage will be low. Since this voltage is applied to the grid of V3, it keeps V3 cut off as long as the background noise is so high that it would drown out any signals (a very low signal-to-noise condition).

In the presence of a signal the background noise drops off to some extent. This drop reduces the conduction of V2 so that its plate voltage increases, thus raising the grid of V3 to a point where V3 can conduct and pass the desired audio signals.

F-M RECEIVERS AND NOISE

Squelch circuits in f-m receivers often interfere with the tracking down of noise sources, or even the checking for the presence of noise. Normally f-m receivers will not reproduce amplitude-modulated noise such as ignition noise. This is why f-m units are usually more quiet than a-m. However, strong a-m signals and unmodulated signals at the frequency of the f-m receiver can overload the receiver circuits, particularly the i-f circuits; the net effect is that they become less sensitive. This means that ignition noise can reduce the effectiveness of an f-m receiver, even though the noise is not heard. Therefore, an f-m system may have a noise problem without it being obvious. If the squelch circuit has an override button or control, it is a simple matter to cut the squelch and check receiver sensitivity. If the squelch does not have a cutout control, the problem is more difficult since sensitivity must be judged while a signal is being received. If you are operating mobile and are near the base station, the strong station signal will overcome the noise, and you will not notice the loss of sensitivity. If you make the test away from the base station, the signal will get through and will appear weak. Thus, it is easy to confuse interference noise with a weak receiver or a weak station signal. If you are experiencing weak reception with an f-m system, it is recommended that you check for possible noise with a scope or a signal tracer. Refer to "Locating Interference Sources" in Chapter 3.

CHAPTER 5

Automatic Noise Limiters

Practically all mobile communications receivers incorporate some type of automatic noise limiter in their circuitry. This applies to both auto and marine two-way communication systems. Noise-limiter circuits are usually independent of the squelch circuits discussed in Chapter 4. One circuit, however—the twin noise squelch—combines both features.

It is possible to modify a receiver to incorporate noise-limiter circuits if none are provided. The amateur-radio and CB magazines are filled with articles on building noise-limiter circuits for every make and model of receiver or transceiver. Commercial noise-limiter units can be attached to a receiver or wired into the circuit.

The following paragraphs describe those noise-limiter circuits that are in common use or that the author feels are particularly effective. No attempt has been made to include all circuits or instructions for incorporating such circuits into a particular make or model of receiver. However, this chapter will give you an understanding of how the various circuits function and how they are related to the rest of the receiver.

The purpose of any noise limiter is to reduce the level of signals interfering with the desired communications. Objectionable noise has two characteristics that make it different from intelligible sound—it is shorter in duration and much stronger in volume. There are other types of noise (such as hum) that do not fit this description; they may not be much louder than the desired sound, and they may be continuous. However, these do not originate in the engine; they are usually a result of defective receiver circuits.

Why can noise pulses disrupt normal sound if they are so short in duration? There are three basic reasons. First, your ears cannot adjust rapidly to a change in noise level. A loud bang nearby renders the ear insensitive to normal sounds for a few moments. Similarly radio static makes it impossible to understand what is being said immediately afterward. If there is a series of loud pulses, such as ignition noise (even though the individual pulses may be brief), normal hearing cannot recover between the pulses to pick out the sound.

Second, a speaker continues to ring or vibrate briefly after it has been subjected to a loud signal. It cannot reproduce the desired sound when it is still vibrating from the undesired noise.

Third, the audio circuits of the receiver can be overloaded by a large noise pulse. Coupling capacitors can be fully charged so they take some time to discharge back to their normal operating level. During this time the capacitors cannot pass audio signals in the normal manner.

Noise-limiter circuits are designed to reduce pulse noises—high, narrow spikes of sound. Circuits can be divided into several broad classifications, such as peak limiters and rate-of-change limiters; there is also the twin noise squelch (TNS) that blocks all sound during noise pulses. Peak limiters chop off the tops of the noise pulses so that the noise level is brought down to that of the signal level. They are relatively simple in circuitry and operation. Typical circuits have been in use for a number of years in various forms. Rate-of-change limiters chop out signals such as ignition noise pulses that change polarity rapidly. These circuits are new in comparison to peak limiters and have not yet been proven in all types of receivers.

SERIES LIMITERS

There are two types of series noise limiters—the half-wave limiter and the full-wave limiter. Both circuits chop off the peaks of noise pulses in order to bring the noise level down to about the same level as the strongest probable signal. The full-wave noise limiter operates on both positive and negative noise pulses, while the half-wave limiter restricts the positive peaks only.

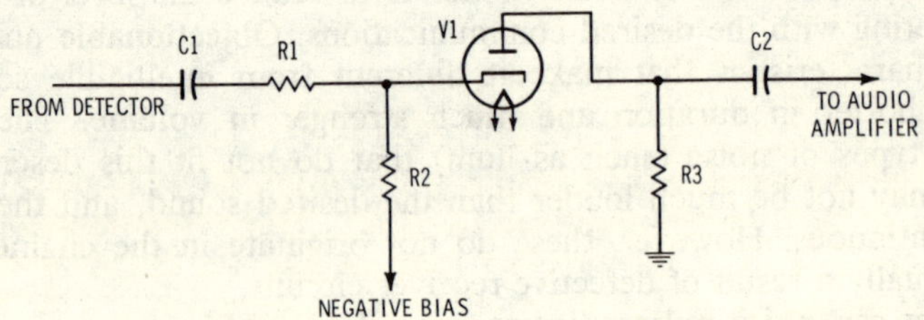

Fig. 5-1. Half-wave series limiter.

A basic half-wave limiter circuit is shown in Fig. 5-1. The cathode is normally biased so that it remains negative with respect to the plate even during positive audio signals from the detector. Under these conditions V1 conducts, allowing the audio signal to pass from the detector to the audio amplifier in the normal manner. A positive noise pulse drives the cathode of V1 positive in respect to the plate. Electrons can no longer flow, so V1 does not conduct the noise pulse to the audio amplifier. A noise pulse is of very short duration. As soon as it is over, the cathode again becomes negative in respect to the plate, tube V1 conducts, and normal operation is restored. The half-wave limiter affects only positive peaks, reducing them to normal levels so they do not interfere with communication.

A basic full-wave limiter circuit is shown in Fig. 5-2. Both cathodes are normally biased so they are negative with respect to their corresponding plates as long as the signal is less than a certain peak-

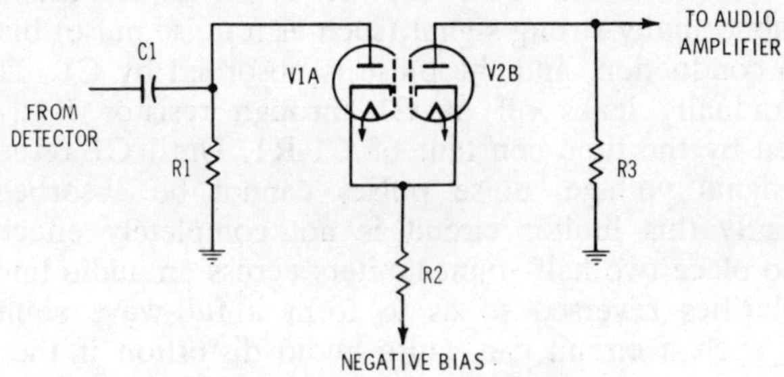

Fig. 5-2. Full-wave series limiter.

to-peak voltage. The audio signal passes from the detector to the audio amplifier without change. When there is a strong negative noise pulse across C1, the plate of V1A is driven negative and it stops conducting, blocking the noise. As soon as the plate is positive with respect to the cathode, V1A again conducts, permitting passage of the audio signals. Positive noise pulses are clipped in a similar manner by V1B since this half of the tube cannot conduct when a noise pulse drives the cathode positive. A combination of the two diodes effectively removes the noise pulses from the circuit and the output sound.

SHUNT LIMITERS

The purpose of a series limiter is to cut off the audio path between detector and audio amplifier; a shunt limiter shorts noise peaks to ground. This is the simplest form of limiting circuit and can be added to almost any receiver. Only two components, a diode and a capacitor, are required.

A basic half-wave shunt limiter is shown in Fig. 5-3. Capacitor C1 charges to the average signal level through CR1. When this point is reached, the voltages on both sides of diode CR1 are approximately equal, and CR1 no longer conducts. The audio signal then passes

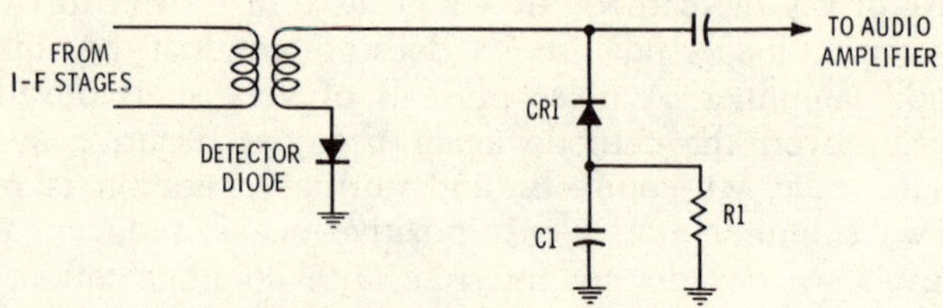

Fig. 5-3. Half-wave shunt limiter.

from the detector to the audio amplifier in the normal manner. However, an abnormally strong signal (such as a noise pulse) biases diode CR1 into conduction, and the pulse is absorbed by C1. This excess charge gradually leaks off of C1 through resistor R1 at a rate determined by the time constant of C1-R1. Until C1 returns to the average signal voltage, noise pulses cannot be absorbed by C1. Consequently this limiter circuit is not completely effective. It is possible to place two half-shunt limiters across an audio line with the diode polarities reversed so as to form a full-wave shunt limiter. However, such a circuit can cause audio distortion if the capacitor values are properly chosen or if the diode characteristics are not correct. As a result, full-wave shunt limiters are rarely used in the audio stages, although they are sometimes used in i-f stages of single-sideband receivers.

THE TWIN NOISE SQUELCH (TNS) CIRCUIT

The twin noise squelch (TNS) circuit has become quite popular in both auto and marine communications receivers; there are many variations in current use. Units are available that can be added to a receiver in order to incorporate an auxiliary TNS circuit. The schematic of such a commercial unit is shown in Fig. 5-4; a photograph appears in Fig. 5-5.

Operation of the circuit is as follows: The output of the receiver detector circuit consists of an audio signal and a d-c voltage that results from the rectification of the carrier. This output is applied across a divider consisting of two 100K resistors and one 20K. The divider is tapped so the grid of V2A receives a larger proportion of the detector output than the grid of V2B. The audio signal is amplified by V2B and goes through gating diodes V1A and V1B before

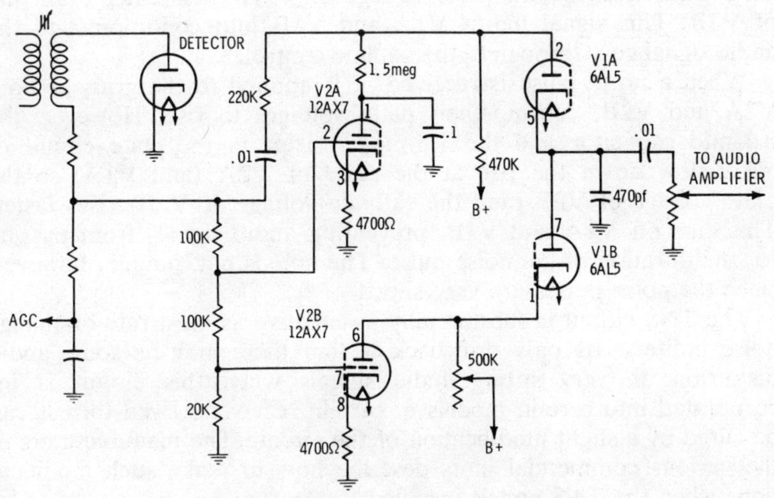

Fig. 5-4. Twin noise squelch circuit.

it reaches the volume control and the audio amplifier of the receiver. A signal can pass through V1B and V1A only when their plates are positive with respect to their cathodes. The charge on the cathode of V1B is controlled by the plate voltage of V2B; the charge on the plate of V1A is controlled by the plate voltage of V2A.

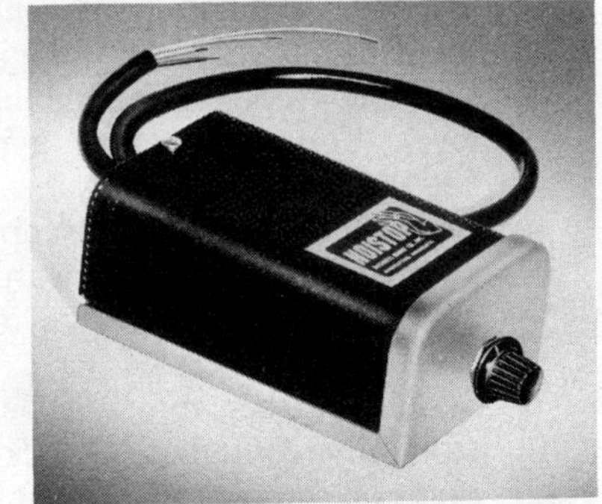

Fig. 5-5. Noise-limiter attachment.

Courtesy Business Radio, Inc.

Under no-signal conditions the grids of V2A and V2B are not biased; both tubes are conducting. The squelch control (supplying B+ to V2B) can be set so the cathode of V1B is sufficiently positive to cut off the gating tubes (V1A and V1B). When there is a signal from the detector, the grid voltage of V2A drops faster than that of

V2B; consequently, the plate voltage of V2A rises faster than that of V2B. This signal biases V1A and V1B into condition, and the audio signal goes through to the volume control.

When a noise pulse is received, it is applied to the grids of both V2A and V2B, causing their plate voltages to rise. However, the 0.1-mfd capacitor and the 1.5-meg resistor in the plate circuit of V2A slow down the rise at the plate of V2A (and V1A) so the plate voltage of V2B (and the cathode voltage of V1B) rises faster. This cuts off V1A and V1B, preventing audio signals from passing for the duration of the noise pulse. The hole is not audible, however, since the noise pulses are very short.

The TNS circuit is substantially as effective as most rate-of-change noise limiters. Its only drawback is that there may be some audio distortion on very strong audio signals when this circuit is incorporated into certain models of certain receivers. Even then it can be cured by a slight modification of the circuit. The manufacturers of the various commercial units describe how to make such modifications when the TNS unit is installed.

Operating the TNS circuit is somewhat tricky the first time you try it, so here is the basic procedure.

1. Turn the receiver on and allow it to warm up for the normal length of time.
2. Turn the TNS squelch control fully counterclockwise. If the receiver has a built-in squelch, open it fully by turning the control to the point of maximum sensitivity.
3. Select a channel where no stations are operating for the moment, or wait until there is a quiet period.
4. Set the receiver volume control to a comfortable listening level. You should hear a rushing sound in your receiver.
5. Set the TNS control by turning it clockwise to a point where the rushing begins to disappear. At this point even the weakest signal may be clearly heard. Ignition noise should be almost entirely absent, but there may still be a slight rushing sound on some receivers.
6. Should further quieting be desired, advance the TNS control only enough to knock out the noise pulses. From this point on the control will act like a very efficient squelch control.

As you become acquainted with the operation of the TNS circuit, you will learn where it is best to set the squelch control for conditions as they exist in your particular area. If you try to reach the threshold point with the volume in the full on position, the weakest noise pulses will be greatly magnified by your volume setting. There may be a tendency to set the TNS squelch too far clockwise, thus taking out some of the weaker signals along with the noise pulses. You will find that by dropping the volume control to a comfortable level, it is not necessary to have your TNS squelch as far clockwise.

I-F NOISE LIMITERS

In sideband-type receivers it is often very helpful to have a limiter that will work on noise pulses before they reach the detector. This is because the beat-frequency oscillator (BFO) used for sideband operation produces an artificial carrier that is much stronger than the actual carrier. If a limiter circuit were to use this BFO carrier as a reference, the limiter would have very little effect on noise pulses. A practical solution is to use a full-wave shunt limiter similar to that previously described, connected across the primary winding of an i-f transformer. Such a limiter, or IFNL, is shown in Fig. 5-6.

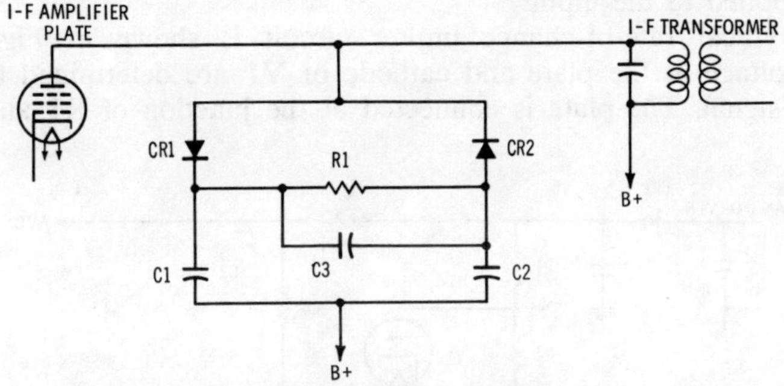

Fig. 5-6. IFNL circuit.

Under normal operation the plate current of an i-f amplifier varies with the incoming signal; stronger signals produce greater variations. In the IFNL circuit capacitors C1 and C2 are charged to the average value of the varying i-f voltage. Since diodes CR1 and CR2 are reversed, capacitors C1 and C2 are charged by both positive and negative swings in i-f plate current. Capacitor C3 and resistor R1 aid the voltage averaging process.

When capacitors C1 and C2 are charged to the average value, the voltage on both sides of diodes CR1 and CR2 is approximately the same, and the diodes do not conduct. This allows the normal plate-current variations to appear across the i-f transformer primary. When there is an abnormally strong signal (such as a noise pulse), the voltage is considerably different from that of capacitors C1 and C2, creating a voltage differential across CR1 and CR2. One of the diodes then conducts and creates a virtual short circuit across the i-f transformer primary. This prevents any signal from passing through the i-f stage. Once the noise pulse drops back to the average i-f level, the voltage differential across the diodes is removed, and the circuit is restored to normal operation.

The IFNL circuit is sometimes used in the i-f stages of an a-m receiver, but its primary use is with single-sideband and double-sideband suppressed-carrier receivers. To be effective the IFNL should be in the first i-f stage.

RATE-OF-CHANGE LIMITERS

Unlike peak limiters, rate-of-change limiters sense the speed at which the instantaneous output voltage is varying, not just the amount of amplitude of the change. In the presence of a rapidly changing output (such as a noise pulse), the rate-of-change limiter remains inactive up to a certain point. As the output swings above that point, this circuit substitutes its own fixed output signal for that applied to the input.

A typical rate-of-change limiter circuit is shown in Fig. 5-7. The voltage at the plate and cathode of V1 are determined by the audio signal. The plate is connected at the junction of R1 and R2,

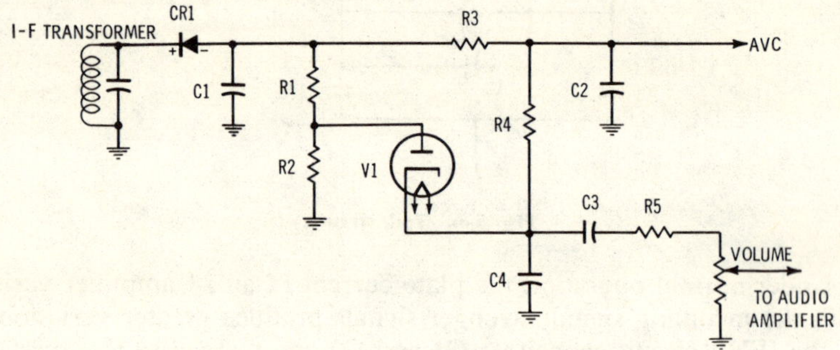

Fig. 5-7. Rate-of-change noise limiter.

and the cathode voltage is taken from the junction of R4 and C4. The values of R1, R2, R4, and C4 are chosen so that the cathode is normally less positive (or more negative) than the plate. This causes V1 to conduct and pass the audio signals from the detector to the volume control. With normal audio variations (no rapid changing of amplitude or polarity) both the plate and cathode follow the audio signals.

When there is a noise pulse or any signal that changes rapidly, the plate of V1 swings instantly in the negative direction. The cathode of V1 also swings in the same direction but not as rapidly as the plate, because of the delay caused by the charging time of C4.

62

Therefore the plate is more negative than the cathode during the noise pulse. This cuts off V1, preventing audio from passing to the volume control. The slow charging voltage of C4 is substituted for the normal audio signals.

When the noise pulse has passed, the plate of V1 returns to the normal voltage. V1 starts conducting immediately, and the audio signals are allowed to pass.

Most rate-of-change limiter circuits obtain their voltage from the detector load resistor (through a filter) and take the audio signal from a tap on this resistor. This makes the rate-of-change limiter self adjusting for various signal strengths.

THE MAKINO AUTOMATIC NOISE LIMITER

The Makino circuit was developed by Japanese engineers for use with communications systems installed on helicopters. (Helicopter engines produce tremendous amounts of ignition noise!) The Makino circuit is similar to a half-wave series limiter, but its effect on noise is more like that of the TNS circuit previously discussed. Instead of chopping off the noise peaks, the Makino circuit punches a hole in the audio each time a noise pulse appears. Since noise pulses are very brief, this hole can not be noticed by ear, but the net effect is that the noise has been removed.

A typical Makino limiter circuit is shown in Fig. 5-8. The detector portion of the Makino circuit consists of diode CR1 and load resistors

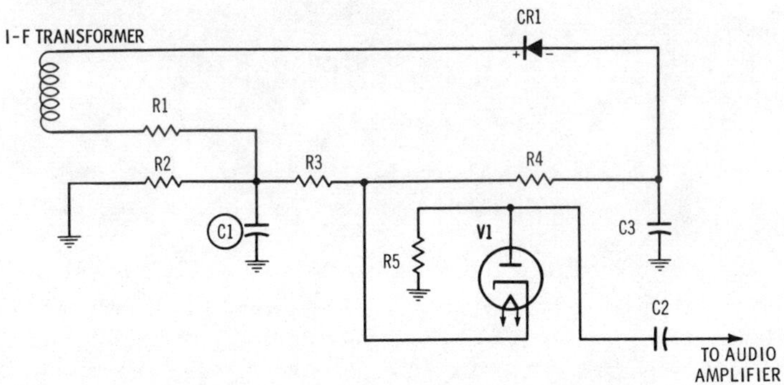

Fig. 5-8. Makino noise-limiter circuit.

R1, R3, and R4. Current through the detector circuit is determined by the polarity of CR1; this diode is connected so the junction of R4 and C3 is the most negative. Audio at this end of the load is shorted to ground through C3. The positive ends of the load resistors are connected to ground through R2. The high resistance of R2 prevents audio from being shorted out, but provides a d-c path to ground. The plate of V1 is connected to ground through R5. As a result, the plate of V1 is normally positive with respect to the cathode, so the tube conducts and passes the audio signals from the detector to the volume control. With normal audio variations both the plate and cathode follow the audio signal.

When there is a noise pulse, voltages on both the cathode and plate of V1 rise. However, C3 holds the d-c level almost constant, so the cathode of V1 goes positive for the duration of the pulse. This cuts off V1 and prevents audio from passing to the volume control. When the noise pulse has passed, the cathode of V1 returns to its normal voltage. This causes V1 to start conducting immediately, and the audio signals are allowed to pass.

CHAPTER 6

Miscellaneous Interference Problems

The engine ignition system and the electrical system are not the only sources of electrical interference. Automobile or marine instruments, automobile wheels and tires, turn and stop signals, power-supply vibrators, and the antenna can cause various degrees of interference. In many cases filtering eliminates the problem. However, because of the nature of these noise sources, they may require special techniques. Marine installations very often need special components and methods to suppress engine interference.

FUEL-GAUGE NOISE

Two points in the fuel-gauge circuits of late-model autos may create electrical noise. These can be checked with the engine switched off and the ignition switch or key in the accessory position.

One source of fuel-gauge noise is the sending unit in the gas tank. This usually consists of a variable resistor operated by a float mechanism. It is mounted within the fuel tank and is usually accessible from the top of the tank. On most U.S. autos this means that the sending unit can be reached through an access plate in the floor of the trunk. Fig. 6-1 shows how the wire from the sending unit is routed along the flooring to the frame, where it is cabled with other wiring going to the dash panel. The easiest way to check for noise from this source is to push the rear bumper up and down. This has the same effect as raising or lowering the fuel level and causes the float to move up and down. Poor contact spots on the variable resistor will cause arcing (and noise) to be conducted back along the

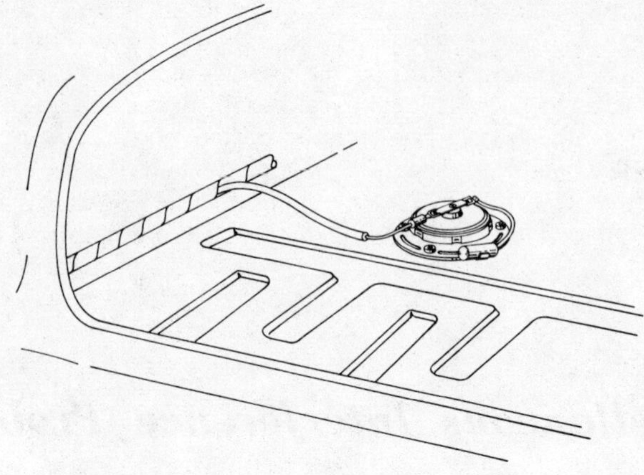

Fig. 6-1. Bypassing a fuel-gauge sending unit.

power line to the receiver. Except in extreme cases, a bypass capacitor connected from the sending unit to ground will cure the problem. The value should be between 0.5 and 1.0 mfd. The electrical connections for such a capacitor are shown in Fig. 6-2; a typical arrangement is shown in Fig. 6-1.

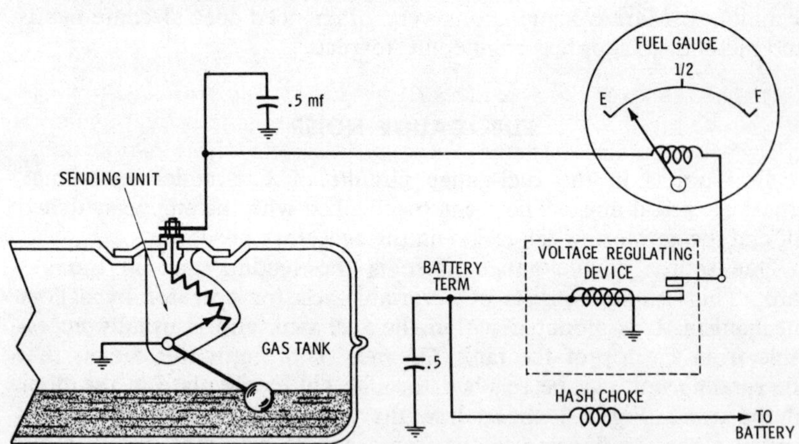

Fig. 6-2. Typical fuel-gauge wiring.

In some cases it is possible to connect the bypass capacitor at the fuel gauge rather than at the sending unit. However, such a connection is not so effective. The sending unit on most fuel tanks is provided with a cover that must be replaced and resealed to prevent exhaust fumes from entering the auto.

Many autos now have a voltage regulator for the instrument cluster. Such a regulator drops the battery/generator voltage from about 12 volts to 5 or 6 volts. The regulation is necessary to keep the various gauge readings accurate despite variations in battery/generator voltage. This device is independent of the regular voltage regulator that controls generator output. It operates by means of a bimetallic arm in a heating coil. When the ignition switch is turned on, the heating coil heats the arm, causing it to bend, which opens the coil contacts. This disconnects the supply voltage from the coil, allowing the bimetallic element to cool and bend back. When the contacts close, the entire operation is repeated. The result is a pulsating output voltage that has a constant average or effective value and is independent of battery voltage. As the battery voltage rises, the contacts make and break more frequently; as the battery voltage goes down, the contacts operate at a slower rate. Thus the output is held constant over all normal input variations.

The opening and closing of the regulator contacts produce "hash" —radio interference—over a wide range of frequencies. Since there are both high- and low-frequency components, both bypass capacitors and hash chokes are required for instrument voltage-regulator noise (Fig. 6-2). If a choke is used in this application, it must be connected as close as possible to the device and not at the battery end of the lead. Also, the bypass-capacitor leads must be kept short to eliminate any tendency to radiate the hash.

Pinpointing noise to the instrument voltage regulator can be a problem. The obvious method is to hit or jar the regulator and listen for noise in the receiver. The catch here is that the regulator is usually located near the instrument cluster and is mounted on the dash panel. When you strike the regulator, you also jar the receiver, so it is not easy to limit the source of trouble to the regulator. A poor connection in the receiver circuits or in the wiring would show similar symptoms. If you experience noise when the dash panel is jarred, try lifting the battery lead to the instrument regulator. Then check again for the same noise condition. Make certain that you switch off the power before disconnecting any power leads. Also, keep the lead away from any exposed portion of the frame. Otherwise, you will have a direct short that can burn out your wiring.

TEMPERATURE-GAUGE NOISE

On rare occasions, the sending unit of the temperature-gauge system may cause noise. This unit is mounted on the engine block and usually receives a stabilized voltage from the same regulator as the fuel gauge. The only practical way to locate temperature-gauge noise is to lift the lead from the sender unit and check whether the noise is eliminated. When the trouble is traced to the sender, a 0.5- or 1.0-mfd bypass capacitor from the temperature gauge itself (not at the sender unit) to ground will usually solve the problem. The sender unit of a temperature gauge consists of a variable resistor in the ground end of the circuit, and, therefore requires only one connection.

WHEEL AND TIRE STATIC

When static electricity is discharged it produces interference, since the discharge involves an electrical spark. Both the wheels and the tires of an automobile are sources of static electricity. However, interference of this type will show up only when the auto is in motion. One way to pin down wheel or tire static is to switch off the engine and let the auto coast down a slight incline. If interference is still present (usually an irregular rushing sound), it is likely that such interference is from the wheels or tires.

In most cases wheel static occurs in the front wheels and is caused by the insulating film produced by the lubricant in the wheel bearings. Commercial collector springs are available to offset this condition (Fig. 6-3). They provide a grounding path from the rotating

Fig. 6-3. Wheel-static collectors.

wheel to the axle. They are mounted within the hub caps and dissipate any static electricity before it can arc (Fig. 6-4). The collector springs are simple to install. They are available in various sizes to accommodate different size hub caps. For small foreign autos it is possible to clip off the ends of the springs to make them fit.

Be careful to follow instructions when installing collector springs. Do not allow the cotter key in the wheel spindle to touch the collector, since this will defeat the purpose.

In some cases static discharge takes place between the auto tires and the road surface. This cannot be eliminated by collector springs but requires anti-static powder in the tires or tubes. Such powder is available in kit form. When properly applied, the powder coats the inside of the tire with a conductive material, thus preventing

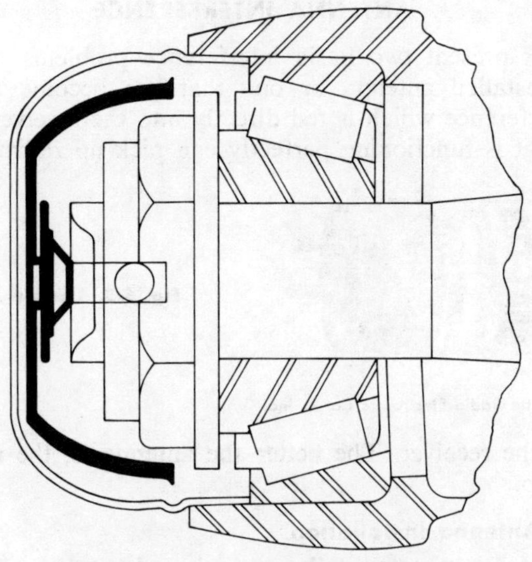

Fig. 6-4. Installed static-electricity collector springs.

static discharge. Tire static can be checked by washing the tire with water, since the water provides a conductive path to ground. Tire static is likely to be encountered during a hot, dry season.

VIBRATOR INTERFERENCE

Since vibrators are mechanical devices that operate on the make and break of contacts, they can produce sparks and be a source of interference. For this reason the vibrator circuits of most power supplies are provided with heavy filtering. However, it is still possible for a vibrator to produce electronic hash under many conditions. One solution to this problem is to replace the mechanical vibrator with a transistor vibrator. These replacements contain no moving parts; yet they provide the same a-c output from the d-c battery supply. Most transistor vibrator replacements contain a multivibrator circuit made up of two or more transistors. The circuit is sealed in a can identical in size and shape to the original mechanical vibrator. No maintenance is required for the replacement vibrators, and installation is quite simple. You simply remove the existing mechanical vibrator and plug the replacement in the same socket. The only possible drawback to the replacement unit is that it may not be able to handle the full current needed for high-power transmitters. Therefore, be sure to check the current rating of the replacement against that of the original. A typical transistor vibrator is shown in Fig. 6-5.

Antennas present two basic interference problems. First, an improperly installed antenna or one that has become defective can create interference which is fed directly into the receiver. Second, an antenna that is functioning perfectly can pick up radiated noise and

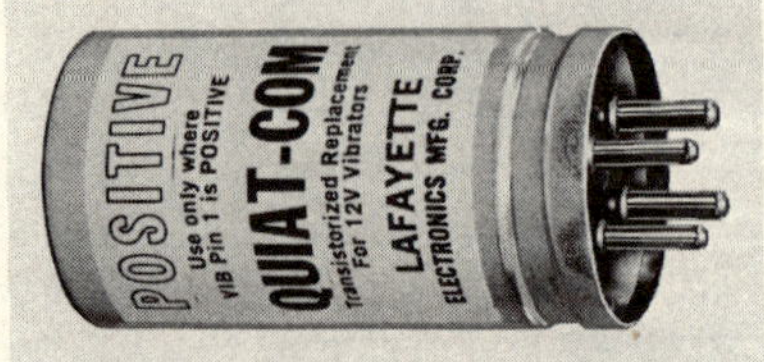

Fig. 6-5. Transistor vibrator.

Courtesy Lafayette Radio Electronics Corp., Inc.

feed it to the receiver. The better the antenna is, the more noise it can pick up.

Checking Antenna Installation

In most cases an automobile or shipboard antenna can be checked with an ohmmeter by using a pair of extra long leads (Fig. 6-6). Connect one ohmmeter lead to the antenna tip and the other to the lead-in that plugs into the receiver. Use alligator clips to make the

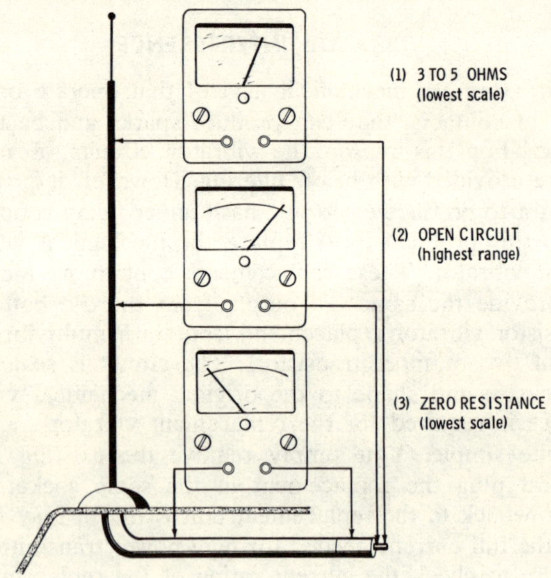

Fig. 6-6. Checking antenna installations.

connection. Set the ohmmeter to its lowest scale. Then shake the antenna and the lead-in. The resistance indicated by the ohmmeter should be 3 to 5 ohms (slightly higher in some antenna systems). If the resistance is considerably higher, or, if it varies when either the antenna or the lead-in is moved, there is a poor connection between the antenna and the lead-in. This reduces antenna efficiency. An antenna connection that is intermittently making and breaking will produce a popping or cracking noise in the receiver when the auto or boat is in motion. In either case, disconnect the lead-in from the antenna and measure the lead-in resistance and the antenna resistance separately. Shake the individual units to see which is causing the problem. Lead-ins are the most likely source of noise, particularly coaxial lead-ins. However, a collapsible antenna can produce noise if there is poor contact between the segments, and an antenna with a built-in loading coil can produce noise if the coil terminals are making poor contact with the antenna base.

If there is continuity from the tip of the antenna to the terminal of the lead-in, connect one ohmmeter lead to the antenna tip and the other to a ground. Again shake the antenna and lead-in. Set to its highest range, the ohmmeter should indicate an open circuit or no contact. If there is a constant high-resistance reading, there is a high-resistance short between the antenna and the auto body or the shipboard ground. Not only will such a condition reduce antenna efficiency but it can be a noise source. If an intermittent short is indicated, a visual check will usually pinpoint the source.

The outer conductor of coaxial lead-ins should also be checked. Connect one ohmmeter lead to the outer conductor at the receiver end and the opposite lead to ground. Set the ohmmeter to its lowest scale and shake the lead-in. The resistance should be zero or near zero. If it is not, the outer conductor is not making proper contact with the ground and is not performing its normal function of shielding the lead-in. Poor shielding offers an easy path for noise to enter your receiver, especially if the lead-in is routed near the engine or near other electrical wiring. In most cases a visual check will pinpoint the poor connection.

Checking for Radiated Noise in an Antenna

Electrical interference from the engine is radiated over a wide range of frequencies. If such interference occurs at the frequency to which the receiver is tuned, a good antenna system will pick it up and transfer it directly to the receiver with maximum efficiency. This does not mean that you should have a poor antenna system, but that you should check radiated noise in your antenna. If the noise is entering the receiver through the antenna rather than through the power wiring, an antenna noise filter may eliminate the problem.

To determine whether the noise is being radiated into the antenna, disconnect the antenna. If the noise is completely eliminated, the problem is one of radiated rather than conducted interference. It is best to check for radiated noise with the antenna trimmer properly adjusted and the receiver tuned to a weak signal. To eliminate error, the regular antenna should be replaced by a dummy antenna that approximates the capacity of the antenna. This brings the antenna coil back into resonance even though the regular antenna is removed. Construction of a dummy antenna is shown in Fig. 6-7.

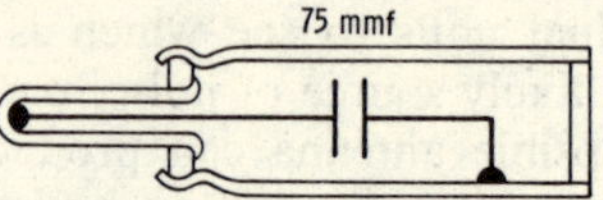

Fig. 6-7. Dummy antenna.

This substitution is done because any noise entering the radio by means other than the antenna is quite likely to be coupled into the antenna input system within the receiver itself. If a dummy antenna is not substituted when the antenna is removed, the input system is de-tuned, making it less sensitive. This gives an erroneous impression of the amount of noise entering the set by means other than the antenna.

If radiated interference is being picked up by the antenna, a a noise filter such as shown in Fig. 6-8 can be inserted into the

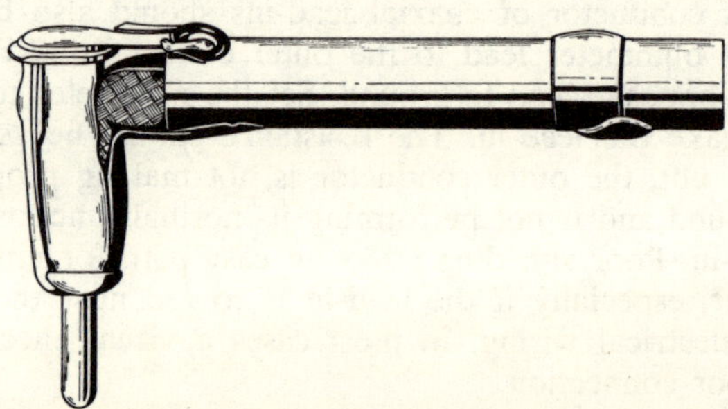

Fig. 6-8. Antenna noise filter.

antenna lead. Such a filter is basically a radio-frequency choke. After it is installed, the antenna trimmer of the receiver or the tuning circuit of the transceiver must be readjusted.

The Location of an Automobile Antenna

The location of an antenna in an automobile installation can be a factor in interference suppression. Generally, a cowl antenna with a short lead to the receiver will transfer the signal with a minimum of loss, as well as provide a minimum of wire for picking up radiated interference. Where there is a particular interference problem, the extra signal strength provided by a short lead can mean the difference between success and failure. There are times when a little extra signal will get through over some unavoidable interference.

In most installations it is best to locate the antenna on the side opposite the ignition wiring. When the wiring tends to be symmetrical (as in most V-8 engines), locate the antenna opposite the distributor. Be sure to remove paint or undercoating from the underside of the mounting surface to assure a good ground connection for the base of the antenna. Also, the fender or cowl on which the antenna is mounted must make good electrical contact with the body and the chassis of the auto. Use copper-braid bonds where necessary.

BONDING SYSTEMS

Bonding provides an easy route for radiated interference to reach ground. The term bonding, as applied here, means connecting two metal objects with a metallic conductor, so that there is a good electrical path between them. For example, both the hood and the frame of an auto are made of metal (except on some sports cars). This metal acts as a shield for containing interference radiated from the engine. However, if the hood is not connected electrically to the frame, the shield is not complete, and interference signals can be radiated through or around the hood. With proper bonding, a way is provided to route the interference from the hood to the frame which acts as a ground.

There are two basic types of bonding: direct bonding and strap bonding. Direct bonding occurs when two metals are connected directly, surface to surface. To be effective the metal surfaces must be in constant contact; they must be physically clean, and they must be unpainted. In most cases it is necessary to join the surfaces with screws or bolts. Bonding can be improved by using toothed washers (internal or external). These washers cut into the metals through any paint or insulating surface; they should be used at both ends of the screws so that both metal surfaces are in contact with the washers and screws.

Strap bonding is accomplished by connecting the two metals with an electrical conductor, usually a braid with lug fasteners at the ends. Braid is used because it is flexible and will not break easily with constant movement. Quite often the two metals to be bonded will move in relation to each other when the auto is in motion. This motion could break an ordinary solid electrical wire. As a side effect, the rubbing of two metals can produce static electricity that creates interference under some conditions. When the parts are bonded together, the static electricity has a discharge path through the strap.

Typical Bonding Points

Some typical bonding points on an automobile installation are:

Corners of the engine to the frame.
The exhaust pipe to the frame and the engine.
Both sides to the hood.
Both side to the trunk lid.
The coil and distributor to the engine and the fire wall.
The air cleaner to the engine block.
Battery ground to the frame.
The generator to the voltage-regulator frame.
Front and rear bumpers to the frames on both sides.
The tail pipe to the frame at the rear.

Engine Bonding

An automobile engine is mounted on rubber shock absorbers that insulate it from the body and the frame. Since the ignition system centers around the engine, the entire engine will radiate ignition noise if it is not grounded. The auto manufacturer grounds the engine in a manner that is satisfactory from an operational standpoint. However, this ground may offer a high impedance to radio frequencies, leaving the engine only partially grounded as far as interference is concerned. When a good radio-frequency ground is provided, the engine acts as a shield and absorbs a good part of the ignition-system radiation. A braid bonding strap can be installed between the engine and the firewall of the auto (Fig. 6-9). This type of bonding is effective and adequate in a great number of installations. In some cases, however, a bonding strap between the engine and chassis will work better, and in extreme cases both may be required.

Hood Bonds

In many instances the hinges do not provide good electrical contact between the hood and the body of the auto. Braid bonds can be used to bypass the hinges and provide the needed ground. Painted surfaces also prevent good electrical contact between the hood and the body, so it is necessary to use metal wipers with sharp points to provide the needed contact. Such wipers are shown in Fig. 6-10. These are often installed at the cowl to contact the underside of the hood and provide good grounding near the antenna. There are various types of hood bonds available for locations where clearance exists between the body and the bond.

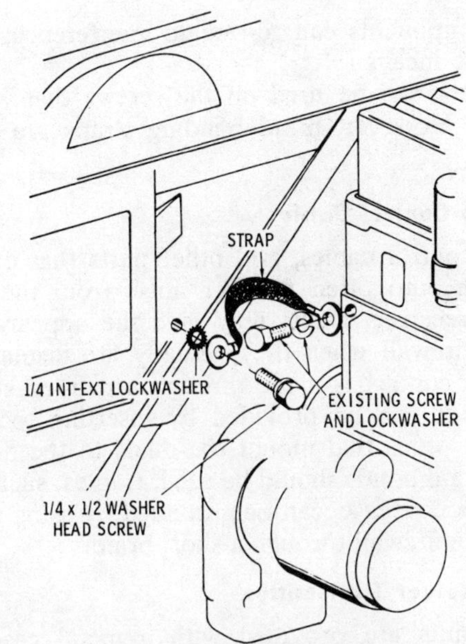

Fig. 6-9. Engine bonding.

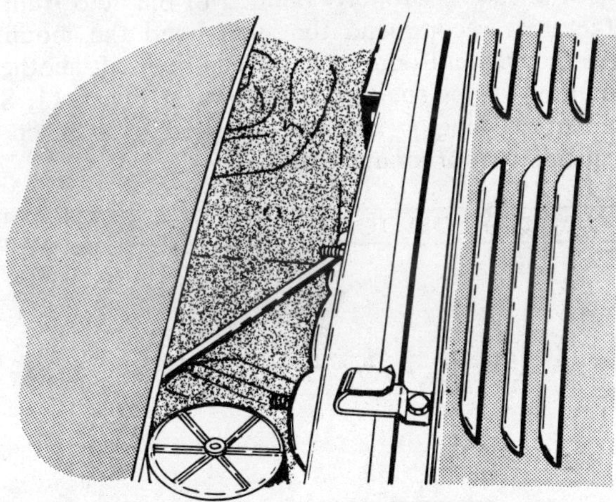

Fig. 6-10. Metal wiper for bonding hoods.

Bonding Body Components

An insulating coating known as anti-squeak or gunk is often used by auto manufacturers to eliminate rattles and squeaks. It is deposited where the body components are joined, and may cause poor grounding of components such as fenders and instrument panels.

These floating components can re-radiate interference; they must be grounded by some means.

Toothed washers can be used on the screws that hold the components together. However, braid bonding straps are usually much more effective.

Heater Ducts and Control Cables

Heater ducts, control cables, and other parts that extend through the firewall of an auto often transfer noise from the engine compartment to the receiver. Ducts may give the appearance of being grounded to the firewall when they actually are insulated or poorly grounded because contact is made through a painted surface. Sometimes good grounds can be provided by inserting toothed washers under the bolts or studs that mount the ducts to the firewall. If this is not practical, braid bonds should be used. Cables, such as the choke control and the hood release, can be grounded by using a cable clamp that connects to the firewall through a short braid.

Mounting the Receiver Transmitter

Most mobile units are provided with a metal case that serves as a shield from radiated interference. However, it will be most effective when the case is properly bonded to the auto frame. In fact most manufacturers recommend that you bond the mounting rack of the radio to the frame. Bonding straps can be used; another method is to use mounting straps such as are shown in Fig. 6-11. Straps are available in various lengths and can be used to support the back of a mobile installation or an auto radio.

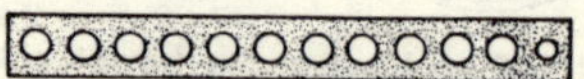

Fig. 6-11. Mounting straps.

MARINE INTERFERENCE PROBLEMS

The same basic techniques used to eliminate engine interference in automobiles are used in boats—suppression, filtering, and shielding. Most marine communications equipment includes automatic

76

noise limiters and squelch circuits, as do automobile units. However, there is one great difference. Neither inboards nor outboards have the built-in metal shielding of auto hoods and bodies, and they do not have an adequate ground. If you succeed in eliminating all conducted noise by means of filtering, radiated noise can still be a problem in your communications system. Noise or interference from a marine engine can also disrupt shipboard depth sounders, radio direction finders, radar, etc., in addition to the marine radiotelephone. Consequently, if you have a serious noise problem on your boat, you should consider shielding the entire system.

One method of shielding is to surround the engine with a copper or bronze screening and connect the screen to the radiotelephone ground plate. Of course this can be used only on inboards. Some marine radiotelephones do not use a ground, so the engine-compartment screen may not be effective. However, no matter what type of engine is used on your boat, there is a shielding system available that covers the entire ignition circuit (distributor, spark plugs, leads, and coil). Details of shielding systems are discussed in Chapter 7.

Many of the outboard manufacturers supply shielding systems for their particular engines. These are usually sold in kit form, and consist of shielded leads, chokes, and spark-plug shields. Typical manufacturer's recommendations are shown in Figs. 6-12 and 6-13. Outboard manufacturers are working steadily to minimize the engine electrical-interference problem, so they may have specific recommendations for unusual and stubborn situations that might be encountered.

Engine instruments and gauges aboard ship should be treated for noise in essentially the same manner as auto instruments. One particular offender in marine engine controls is the tachometer. Noticeable reduction in radio noise level can often be attained by shielding the entire length of the tachometer wire from the control panel to the engine. Since the tach picks up its impulses from the ignition primary, the wiring can radiate these undesired signals like an antenna. Remember to ground the shielded part of the wire to the engine. Some ignition tachometers may require adjustment or may not work at all if shielding is added because the shielding may change the characteristics of the tachometer impulses.

The magneto grounding wires or battery-ignition wires should also be shielded along their entire length from the engine to the control panel; the shielded braid should be grounded. Do not attempt to alter the remote control cables. However, large diameter woven metal braid can be slipped over the electrical cable assembly, and this should be grounded at the remote control and at the engine. On outboards using battery ignition a 0.1-mf feedthrough capacitor installed as close as possible to the first coil (in the switch-coil lead,

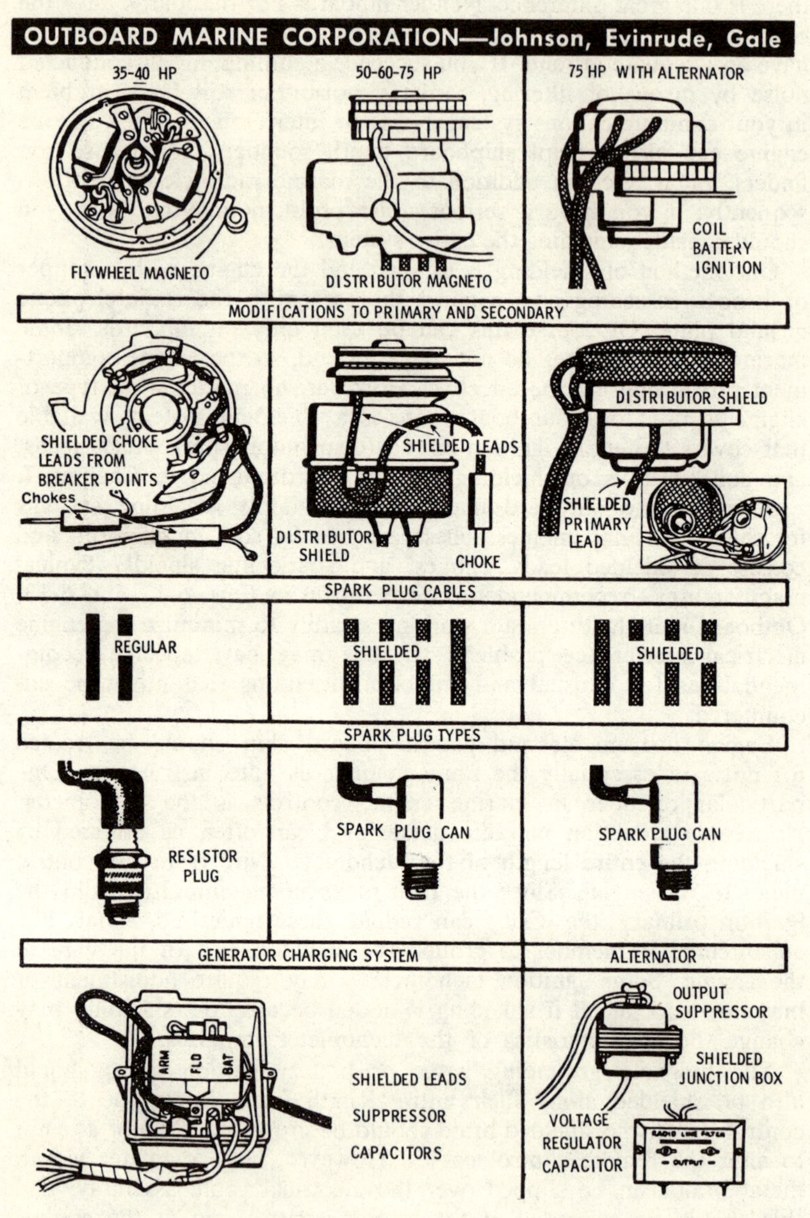

Fig. 6-12. Marine shielding systems.

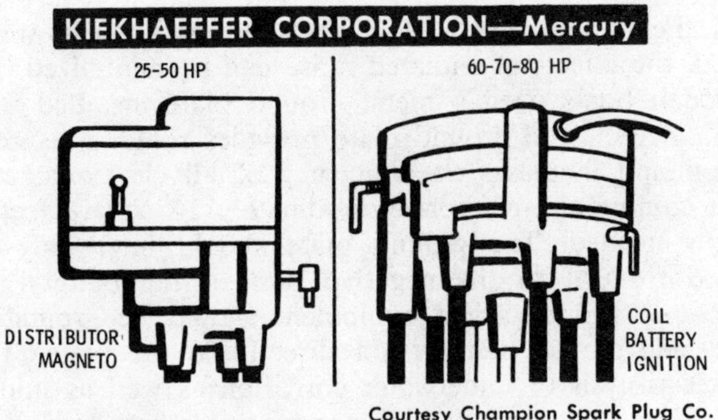

Fig. 6-13. Marine shielding systems.

not the coil-distributor lead) will keep unwanted conducted noise from entering the boat's wiring.

Shielding kits recommended by manufacturers of outboard motors will not affect engine performance—provided the ignition is in top shape. Engines not in good condition may have their problems magnified. Outboards that tend to foul plugs may have greater difficulty when suppression is installed. Before adding any suppression, correct the causes of fouling—excessive combustion-chamber deposits, poor breaker-point synchronization, clogged passages in the manifold bleed-off system, cracks in the distributor cap, improper carburetion, and poor coordination of spark-advance controls with carburetor butterfly movement. Some outboards operated at low speeds (slow troll, for example) may misfire from a carbon build-up between the plug electrodes even though the insulator remains fairly clean. Outboards not usually prone to this type of fouling may become susceptible to gap bridging after suppression is installed. It is good practice to inspect plugs more frequently after modifications have been made. And of course, with or without suppression or shielding, use the proper gasoline and the recommended type and amount of oil.

Rare and seldom checked sources of static noise in a marine system are metal riggings rubbing together, as well as turnbuckles and shackles "sounding off" at certain propeller frequencies. Metals rubbing together can create static electricity; when this discharges, it appears as noise in the receiver. Usually this sound is intermittent and very difficult to locate. However, if you notice a regular crashing sound on the receiver that coincides with ship motion (pitching or rolling), check the metal rigging. After it is located, this type of noise is fairly easy to cure. A bonding strip across the two metals does the trick.

If all the metal cases of the electronic equipment are properly grounded, the effects of radiated noise can be minimized. Fiberglass and wooden boats need a metal ground plate installed underneath the hull since a good ground plate provides some measure of noise protection and increases the efficiency of all electronic equipment. Either a commercial unit or approximately 12 square feet of metal surface is needed. The ground plate should have only one wire connected from it to the negative post of the battery. Likewise, all electronic and electrical equipment should be grounded at the negative side of the battery, not directly to the plate. This helps reduce electrolysis or underwater corrosion as well as minimize the noise problem since both are affected by water flowing against metal.

Metal boats do not require a ground plate if the electrical system is grounded to the hull. All wiring should be routed under the metal rails of the boat if possible to take advantage of the natural shielding effect.

In addition to the procedures previously mentioned, other steps can be taken in the installation and maintenance of marine electronic gear that will minimize the noise problem. These include:

1. Install the antenna on the side opposite from the remote-control engine panel, and as far away from the engine compartment as possible. On an outboard install the antenna well forward. This will keep radiated engine noise from reaching the antenna.
2. Always use a marine-type antenna, not an auto antenna. Make certain that the antenna has the proper electrical characteristics to match the equipment with which it is being used.
3. Use a shielded antenna lead-in wire, and ground the metal braid to the receiver. Many marine radiotelephones are now using coaxial cable, which greatly reduces the pickup of radiated engine noise.
4. Make sure that no wires run through a leaky bilge. For your own protection, do not handle electronic equipment, especially a transmitter, when standing in bilge water or wearing water-soaked shoes. It is best to disconnect the hot head or remove the line fuse when working on electrical equipment.
5. Interference from the small electric motors in windshield wipers, bait tanks, bilge pumps, etc., can be eliminated by installing a 0.5- or 1.0-mfd bypass capacitor from the hot lead to ground. The capacitor should be installed as near to the offending device as possible.

CHAPTER 7

Shielding Systems

Shielded ignition systems offer the most effective method of noise elimination available today. About the only means of combating extreme radiated-noise problems is to shield the ignition wiring, especially where there is a minimum of natural shielding. Two classic examples of this are on motorboats and sports cars with *Fiberglas* bodies. In either of these cases, a completely shielded ignition has the same effect as surrounding the engine with a metal shield. Radiated noise is confined and grounded. Conducted noise associated with the radiated noise can be eliminated with bypass capacitors and filtering.

There are three basic types of shielding—direct-shielded cable with a shielded plug, conduit-shielded cable with a shielded plug, and direct-shielded cable with a plug shield and an unshielded plug (Fig. 7-1). The first two types of shielding are usually installed as original equipment. The third type is installed as part of a kit or as a do-it-yourself system.

The author cannot recommend any of the shielding systems found in the do-it-yourself articles of the many electronics magazines. These systems, although they may provide some measure of shielding (some of them work quite well in that respect), may also impair operation of the ignition. When an ignition system is shielded with any type of cable shield, the voltage produced by the ignition coil is reduced at the spark plugs. The commercial shielding systems are designed to offset this condition. They use insulating materials that will not lose their dielectric properties, even under the extreme heat of the average engine. The home-made systems usually ignore this factor completely.

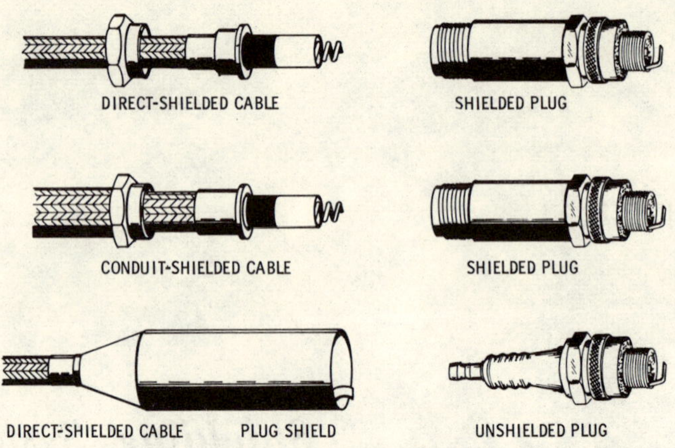

Fig. 7-1. Types of shielding.

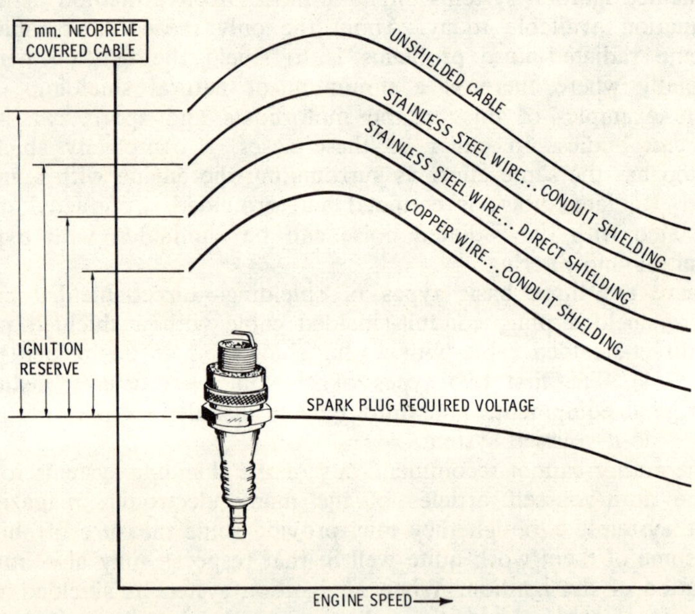

Fig. 7-2. Effect of shielding on ignition systems.

Even with the commercial shielding systems, the ignition must be in good shape—with plenty of reserve voltage available to the plugs. A borderline ignition should not be shielded. The effects of shielding on available coil voltage are shown in Fig. 7-2. The conditions which cause a borderline ignition system can often create noise (see Chapter 1), so the first step in eliminating noise is to put the ignition system in peak working order.

Some of the manufacturers of ignition shielding supply a replacement ignition coil with their kits, particularly for the fully shielded marine systems. These coils supply an increased voltage for the spark plugs. While this is not necessary for all ignition systems, it is a point to remember. If you install a shielding system that cures your noise problem but creates an ignition problem, a new, high-performance coil may be the answer.

SHIELDING THE SPARK PLUGS

There are two alternatives when shielding the spark plugs—external spark-plug shields, and shielded plugs with a harness. The spark-plug shields with direct shielded cables (Fig. 7-3) combine

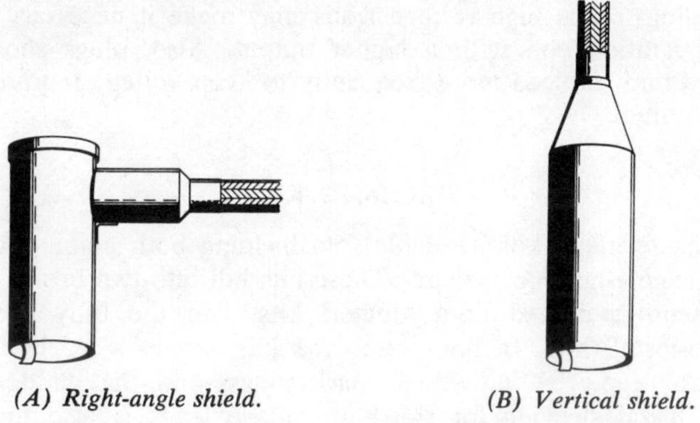

(A) Right-angle shield.   (B) Vertical shield.

Fig. 7-3. Spark-plug shields with direct-shielded cables.

good performance with lower cost; most shielding kits use this system. Regular, resistor, or auxiliary-gap spark plugs can be used. The right-angle plug shield offers low profile installation advantages, but it cannot be used in all applications. The spark-plug shield contains an inside ground spring that must contact the shell of the spark plug. This provides a path to engine ground for unwanted radio interference.

The shielded plugs with a shielded harness (Fig. 7-4) are usually supplied as original equipment on aircraft and marine installations. Before converting to such a system, make sure that there is adequate clearance to accommodate the overall height of the shielded plugs

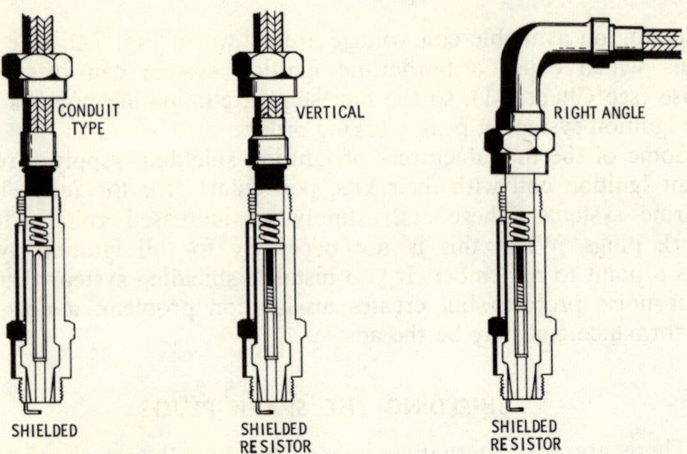

Fig. 7-4. Shielded plugs with shielded harness.

and connectors. When conduit-type shielding is used, avoid the use of cambric-covered or plastic-coated high-voltage leads. The addition of shielding to the high-voltage leads may make it necessary to install an ignition coil with a higher output. Also, plugs should be regapped and serviced more frequently to keep voltage requirements at a minimum.

SHIELDING KITS

There are many kits available for shielding both automobile and marine engine-ignition systems. These kits fall into two broad groups —the semi-assembled do-it-yourself kits, and the fully-assembled ready-to-install kits. In both cases the kits supply a shield for the distributor cap, the individual spark plugs, and the ignition coil; flexible metal sleeving for the high-voltage leads is also included. The difference in the kits lies in the manner of assembly. The semi-assembled system requires that you cut the spark-plug wires to the proper length and attach the shielded distributor cap, the spark-plug shields, and the coil shields. Sometimes existing wiring is used, and the braid is passed over the wires. In the fully-assembled units the old ignition wiring is removed, and pre-assembled items are installed.

The following paragraphs describe the steps necessary for the installation of three typical ignition-shielding systems. By studying these steps it is possible to become familiar with the sources of radiated ignition noise and how to surround them with shielding. Most shielding systems also incorporate some form of filtering to eliminate conducted noise.

Hallett Shielding Kit

The Hallett Manufacturing Company produces a shielding system sold in kit form. It includes shielding for the ignition wiring, spark plugs, distributor, and ignition coil. Alternate kits are available to provide shielding for the round automotive-type distributor and the flat marine-type. The kits come in two basic configurations. On one the shielded ignition wiring must be assembled to the distributor shield, and the old distributor cap is used; the other configuration provides for a completely shielded distributor cap and factory-assembled ignition wiring. It is sold as a one-piece item. A typical installed shielding system is shown in Fig. 7-5.

Courtesy Hallett Manufacturing Co.

Fig. 7-5. Installed shielding system.

NOTE

Always disconnect the battery before doing any work on the ignition system of any engine.

To install a Hallett shielding kit, proceed as follows:

1. Check the distributor cap to see that it is keyed so that it fits on the bowl in only one position. If it is not keyed, mark a reference point on the distributor and the cap. Make a record

of the wiring from the cap to the spark plugs. Remove the wires completely, unfasten the spring clips as necessary, and remove the distributor cap to a bench or other work area.

If your kit is of the type in which the distributor shield and the ignition wiring are supplied as one piece, proceed to Step 11. If your kit is of the partially assembled type, it is necessary to assemble the shield to the distributor cap and the ignition wiring to the shield. For pre-installation assembly of round distributor shields, proceed to Step 2; for pre-installation assembly of the flat, rectangular, marine-type distributor shields go to Step 5.

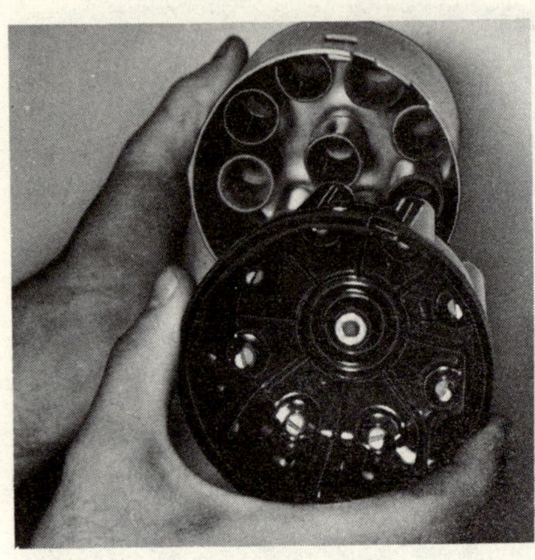

Fig. 7-6. Inserting distributor cap into shield.

2. Place the distributor shield over the cap, making certain that the cap towers line up with the rubber insulators within the shield (Fig. 7-6). The cap towers must fit snugly within the corresponding insulators. Certain caps have cutout areas on the bottom edge to provide for clearance on the distributor bowl. A quick check will indicate whether these are needed for your engine. If they are, align the shield over the cap as necessary to use the cutouts to best advantage. The shield supplied with the kit should have sufficient cutouts to provide clearance on the engine; however, all the cutouts may not be necessary for all engines.

3. The spark-plug wires are numbered, and they must be inserted in the same rotation as the original wiring. Install the shielded wires by pulling approximately 3 inches of black wire out of the shielding braid and inserting it through the insulators in the shield top (Fig. 7-7). Continue to push the cable down until it snaps firmly in place.

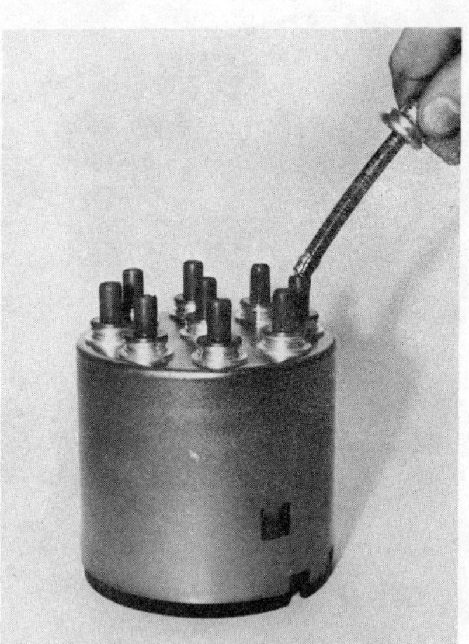

Fig. 7-7. Connecting a cable
to the distributor.

4. Slide the braid and snapnut down to the shield (Fig. 7-8). It
    may be necessary to start at the plug end and work the braid
    down over the inner wire. Using a firm thumb pressure, rock
    the braid snapnut down until it engages the shield fitting. Re-

Fig. 7-8. Connecting the shield
to the distributor.

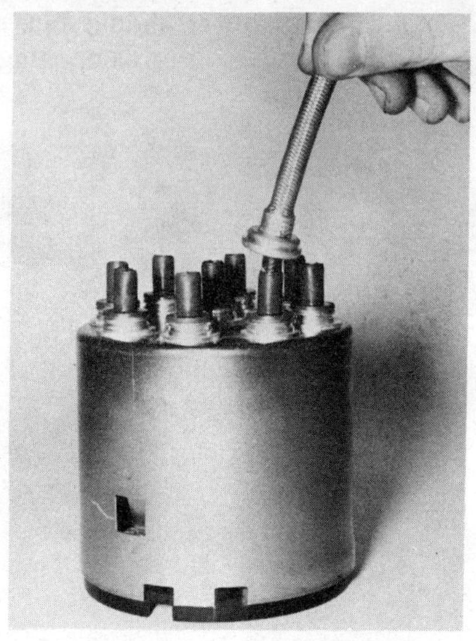

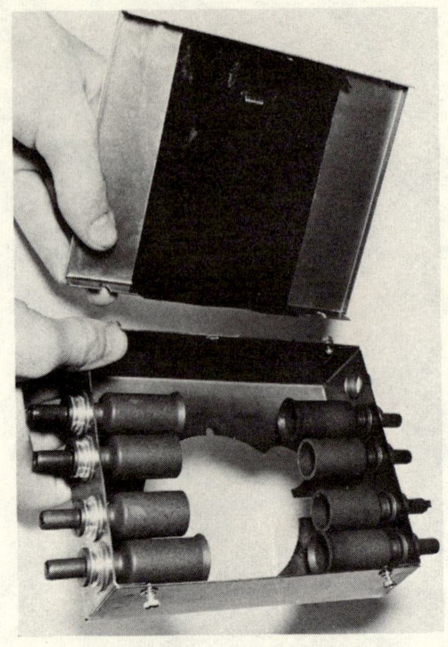

Fig. 7-9. Removing the cover from a marine-distributor shield.

peat this procedure for all wiring, including the center coil-to-distributor wire. Proceed to step 11.

5. For the rectangular, marine-type distributors, remove the cover from the distributor cover (Fig. 7-9).

6. Install the distributor cap in the shield. Insert the four cap towers into the corresponding shield insulators (Fig. 7-10). The cap coil tower should face the opening without an insulator on the side with five openings.

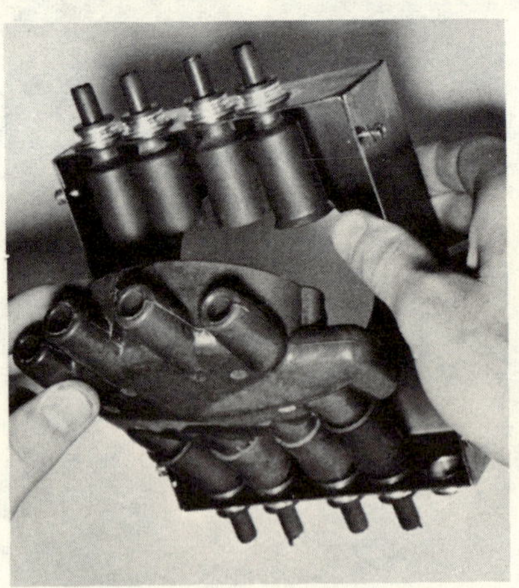

Fig. 7-10. Assembling marine-distributor shield—step one.

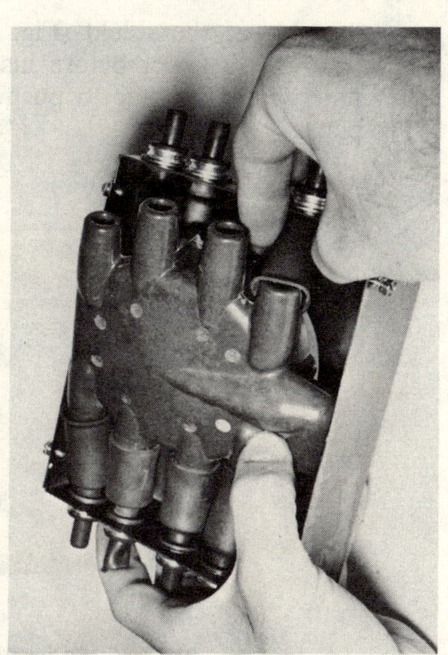

Fig. 7-11. Assembling marine-
distributor shields—step two.

7. Press the cap down, and work each of the remaining four
   shield insulators on the opposite side onto the corresponding
   cap towers (Fig. 7-11).
8. The spark-plug wires are numbered and must be inserted in
   the same rotation as the original wiring. Install the shielded
   wires by pulling approximately 3 inches of black wire out of
   the shielding braid and inserting it through the insulators on

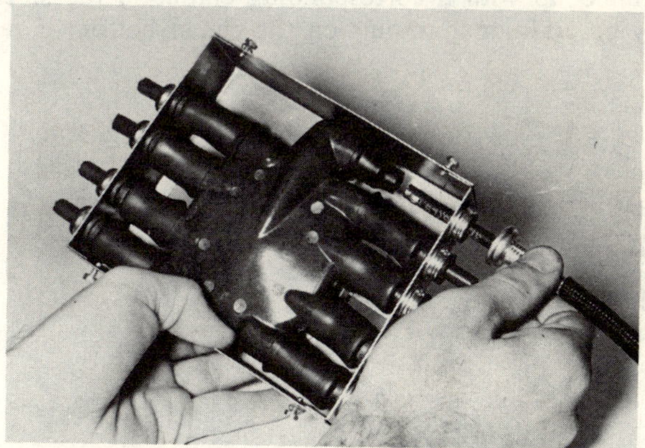

Fig. 7-12. Inserting the coil cable into the distributor cap.

the side of the shield (Fig. 7-12). Install the black insulator on the coil tower before inserting the shielded coil-to-distributor wire. Continue to push the cable until it snaps firmly in place.

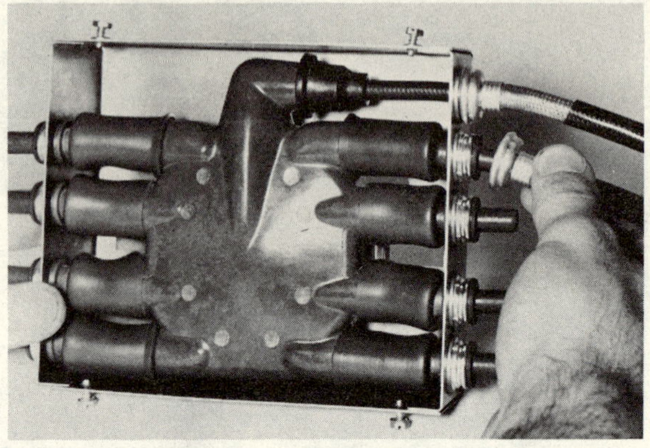

Fig. 7-13. Connecting shielded cable.

9. Slide the braid and snapnut down to the shield (Fig. 7-13), working the braid down over the inner wire. Using a firm thumb pressure, rock the braid snapnut down until it engages the shield fitting. Repeat this procedure for all wiring, including the coil wire.
10. Replace the shield cover and tighten the screws (Fig. 7-14). Check the bottom of the shield and cap (Fig. 7-15). There are two cutouts for the cap snap fasteners. The cap notch should be lined up with a corresponding cutout in the shield. There may be additional cutouts on the shield bottom if required to

Fig. 7-14. Replacing the shield cover.

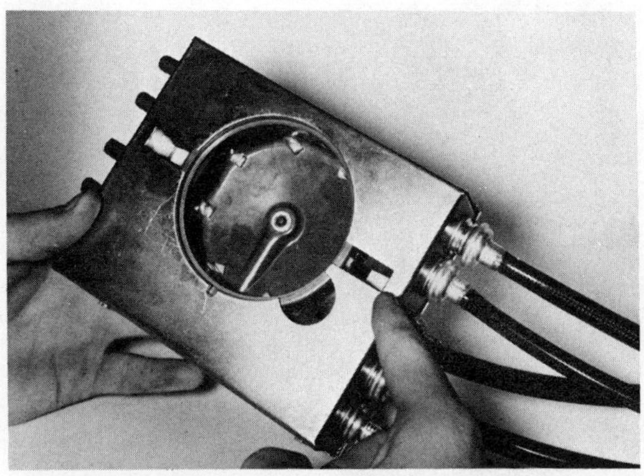

Fig. 7-15. Checking distributor-shield reference points.

provide clearance on your engine. All cutouts may not be necessary for all engines.

11. Place the assembled distributor cap and attached wiring on the distributor. Route the shielded wires to the spark plugs over the same path as the original wiring if possible (Fig. 7-16).

The remaining steps involve installing a shielded ignition coil. For the installation of a shield on an existing coil proceed now to the next step (12). Some kits, particularly those for marine engines, are supplied with a high-performance coil with a built-in shield to be used in place of the original coil (Fig. 7-17). Such a coil provides improved ignition performance and shielding. Since it is normally connected for negative ground, move the wire inside the shield to the opposite coil terminal for engines with a positive battery ground. Also determine if the original coil has a resistor bypass circuit during

Fig. 7-16. Routing shielded ignition wiring to the spark plug.

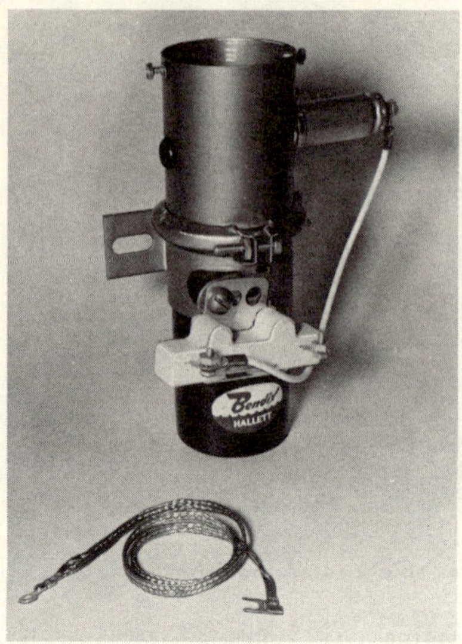

Fig. 7-17. Shielded ignition coil.

the starting cycle. If the engine is not equipped for bypass, a 12-volt relay, actuated from the starter switch, should be added. Never wire around the new coil resistor for continuous operation. If the original system has a series coil resistor, it must be removed or shunted out.

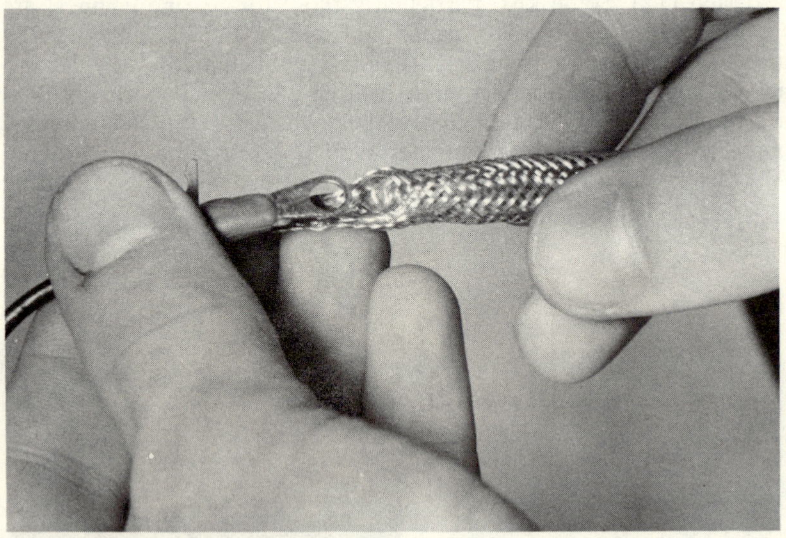

Fig. 7-18. Installing shielded braid on low-tension lead.

12. Determine by examination if the coil can remain in the original
location and still have adequate clearance after the coil shield
is installed. In most cases the original location is satisfactory.
Sometimes it is only necessary to slide or rotate the coil in its
mounting clamp. When the proper location is determined, note
the terminal to which the low-tension lead from the distributor
is connected, then remove the coil and its clamp from the
engine. Scrape paint from the clamp, the engine block, and
the body of the coil where they join. Also scrape paint from
the point where the new coil-shield retaining clamp will fit
on the upper flange of the coil. Return the coil and clamp
to the engine, and bolt in place, since it is not necessary to
remove the coil or distributor from the engine to perform
the following steps.
13. Slide shielded braid from the kit over the end of the low-tension
lead from the distributor (Fig. 7-18). If the lead terminal is
similar to that shown, it need not be replaced. A spade or fork
lug should be replaced with a lug from the kit so that it will
pass through the braid without catching.
14. Work the braid on the lead so that the spade lug ends up at the
distributor. Fasten the lug to the distributor with a screw (Fig.
7-19).
15. Pass the coil end of the low-tension lead through the coil-shield
grommet, and connect it to the correct terminal on the coil
(Fig. 7-20). Connect the lead from the feedthrough capacitor
(inside the shield) to the opposite coil terminal (originally
connected to the ignition switch). Fasten the braid lug (ground)
to the coil shield with the screw provided.

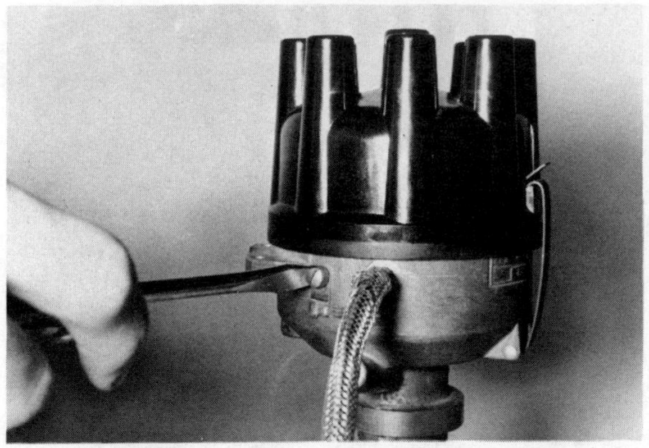

Fig. 7-19. Grounding low-tension lead shielding.

Fig. 7-20. Connecting ignition-coil leads.

16. Place the coil shield on the coil. Rotate the shield so that the feedthrough capacitor is free of any obstructions (Fig. 7-21). Install the coil-shield clamp, and tighten it with a screwdriver. Be sure that the shield and clamp fit snugly over the coil lip. If the coil is bent or out-of-round, it may have to be straightened.

17. Slip the rubber insulator over the top of the coil, and install the cover on the coil shield (Fig. 7-22), taking care that the shield high-tension contact enters the center tower of the coil. Check that the low-tension leads inside the shield are clear of the high-voltage contact. Fasten the shield cover securely to the shield with screws.

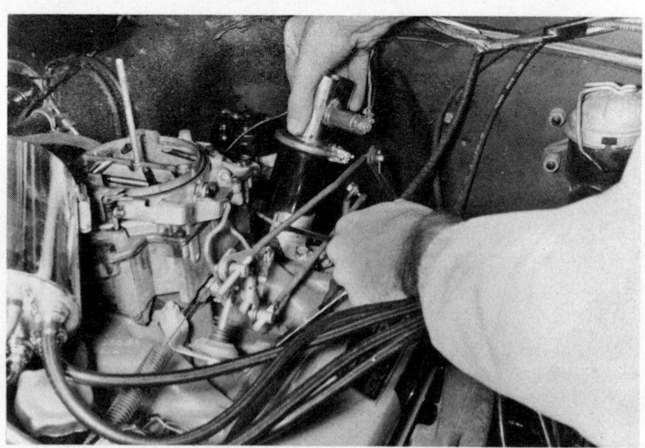

Fig. 7-21. Rotating ignition-coil shield to provide clearance.

Fig. 7-22. Installing ignition-coil shield cover.

18. Put the shielded high-tension lead in the center tower of the distributor. Connect the ignition-switch lead to the coil-shield feedthrough capacitor (Fig. 7-23). If the capacitor is already attached to a series regulator (provided with some new coils), connect the switch lead to the other terminal of the regulator.

Estes Shielding Kit

The Estes Engineering Company produces shielding systems that are sold in kit form. They include shielding for the ignition wiring, spark plugs, distributor, and ignition coil. Variations are available to provide for the round automotive and the flat marine distributors.

Fig. 7-23. Connecting ignition-switch lead.

The kits come in various configurations, both partially assembled and fully assembled. A typical installed ignition-shielding system is shown in Fig. 7-24. A shield with built-in feedthrough capacitors is also available for the voltage regulator (Fig. 7-25).

Courtesy Estes Engineering Co.

7-24. Installed shielding system.

To install an Estes shielding kit, proceed as follows:

1. Before removing any wiring from the engine, make sketches and notes of the firing order, cylinder locations, polarity of the coil primary, and the location of the distributor-cap keyway. A typical sketch might look something like that shown in Fig. 7-26.
2. Disconnect one cable at the battery. This eliminates any chance of accidental damage to engine components during installation.
3. Remove the original distributor cap and spark-plug harness as a unit.
4. Remove the ignition coil. Discard any capacitor connected to the coil.

96

Courtesy Estes Engineering Co.

Fig. 7-25. Voltage-regulator shield.

5. Temporarily install the shielded distributor cap. Make certain the locating keyway in the cap is aligned with the mating locator in the distributor bowl. Check the following:

   a. The shielded high-voltage cable between the distributor and the coil must enter the distributor from the proper direction. If it does not, interchange the cable and the hole plug on the opposite side of the distributor. Ground the shielded cable as it was originally.

   b. The ignition coil must be within reach of the cable from the distributor shield. Some slack should exist in this cable so the distributor cap can be removed in the future during routine servicing. Occasional installations may require a relocation of the coil. If it is mounted on the auto firewall, remount it on the engine block if possible, since this will give better grounding and more effective shielding. If firewall mounting cannot be changed, install a heavy bonding

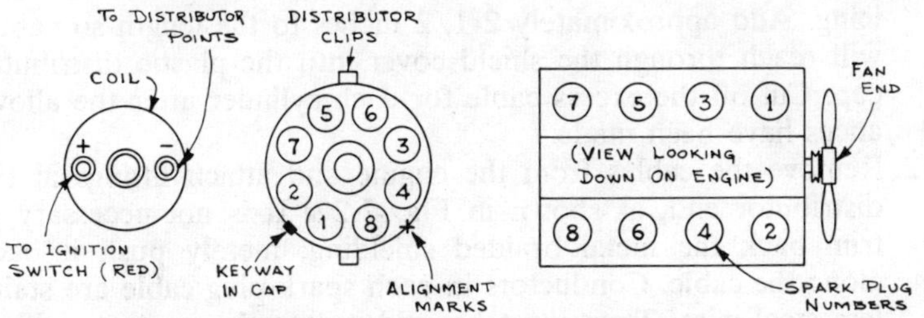

Fig. 7-26. Typical ignition-wiring identification sketch.

jumper of braided shielding or number 10 AWG stranded wire between the firewall and the engine, attaching both ends with tooth lockwashers from the kit.

6. Place the cover on the distributor shield, and align the side cover holes with the smaller mating holes in the shield. Do not install the screws as yet. Refer to the sketch of the original wiring, and mark the cylinder numbers beside the correct holes in the cover. Make a pencil mark on the side for reference in aligning the shield cover and body during final assembly, and note the location of this mark on your sketch.

7. Determine the proper location for the grounding cable—one of the two small holes provided on opposite sides of the distributor shield. Sufficient slack should remain in this cable so that the distributor cap can be removed in the future during routine servicing.

8. Snap each spark-plug shield over a spark plug. The lower skirt of the shield should cover the hex portion of the plug. On engines where the plugs are partially hidden, remove a plug and hold it in your hand while installing it several times in a loose shield. This will give you the proper feel of a correct installation. A slight twisting or rocking action on the shield sometimes helps installation on the plugs. The flat grounding spring inside the shield should press against the flat side of the spark-plug hex.

9. Attach the numbered tapes supplied with the kit to the corresponding spark-plug cable, approximately 3 inches from the spark-plug end. This provides identification during final assembly of the cables to the distributor cap, so be sure that is done carefully and permanently.

10. Group each bank of cables together neatly and route them toward the distributor shield. Remember each group of cables will be supported and grounded at approximately the midpoint by a large circular clamp from the kit.

11. Allow enough slack in the spark-plug cables so that the distributor cap can be removed in the future during routine servicing. Add approximately 2 1/2 inches to the length so cables will reach through the shield cover into the plastic distributor cap. Cut off the excess cable for each cylinder after the allowances have been made.

12. Remove the cables from the engine and attach fittings at the distributor end, as shown in Fig. 7-27. It is not necessary to trim back the metal braided shielding; merely push it back along the cable. Conductors in each spark-plug cable are stainless steel wire. They must be soldered to the contacts with a special soldering flux.

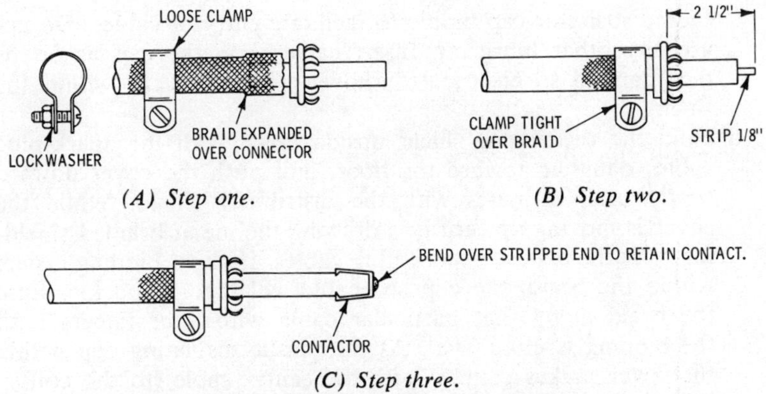

(A) Step one.   (B) Step two.

(C) Step three.

Fig. 7-27. Installing cable fittings.

13. Remove the distributor shield and cover from the engine, and place them on a workbench. Remove the loose cover from the shield. Remember that the relationship of the distributor-cap keyway to the spark-plug cables must agree with the sketch of the original harness. Keep that sketch in front of you during this phase of the installation.

14. Snap the numbered spark-plug cables into the cover, making certain that they are correctly oriented. The installation should now look similar to Fig. 7-28.

15. Pull approximately 10 inches of wire out of the braided shielding. Lightly squeeze the brass contactors into the ignition cable. A slight amount of silicone grease DC-4 has been placed in

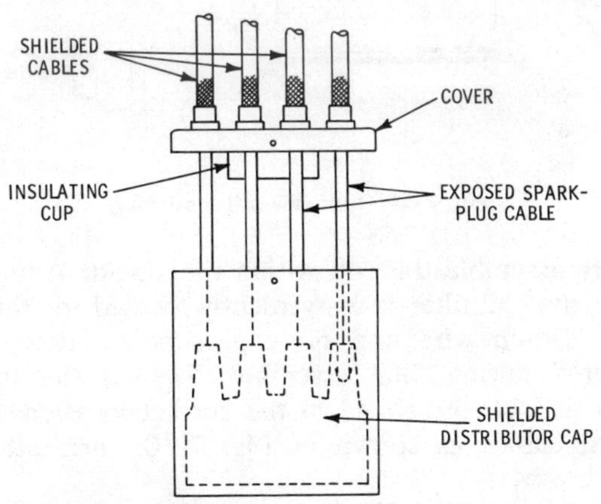

Fig. 7-28. Installing cables in the distributor shield.

each distributor-cap nipple to facilitate entry of cables. Do not use any other lubricant. Insert cables into the cap as far as they can go so electrical flashover will not occur within the shield.

16. Hold the distributor shield upside down with the spark-plug cables dangling toward the floor, and push the cover upward gently until it mates with the distributor shield. While the cover is moving upward it is drawing the metal braided shielding over the exposed spark-plug cables. If slight binding occurs within the braid, the exposed cable will tend to buckle. Ease the braid along that particular cable with your fingers until the binding is eliminated. As the plastic insulating cup within the cover makes contact with the center cable (to the coil) a slight amount of resistance will be encountered. The insulating cup is soft and will flex. When the cover is on all the way, fasten it with screws.

17. Attach the grounding cable to the distributor shield at one of the small holes in the side.

18. Use a piece of sandpaper or a wire wheel to remove all paint and rust from the areas indicated by the arrows in Fig. 7-29. Also remove paint and rust from the engine area where the coil bracket is attached.

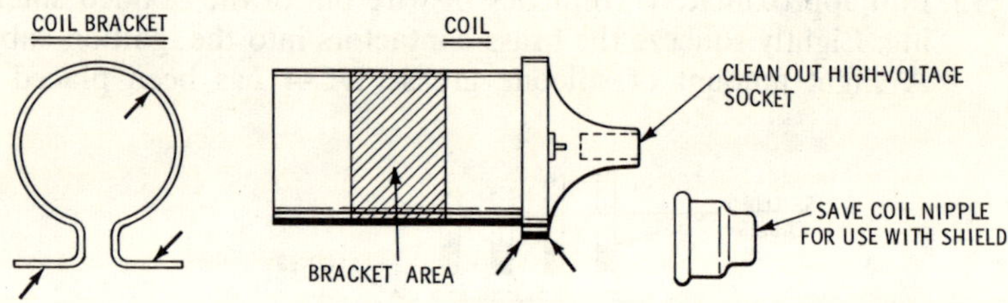

Fig. 7-29. Ignition-coil mounting.

19. Loosely assemble the coil within the shield. Adjust the shield so that the coil filter is conveniently located for the installation of the switch wire and no engine obstructions will be encountered during final assembly. Tighten the three internal screws to lock the shield in the correct location. Connect the internal cables as shown in Fig. 7-30, then attach the coil shield cover.

20. Attach the shielded distributor cap to the engine in the same manner as the original cap. Be sure that the locating tab of the

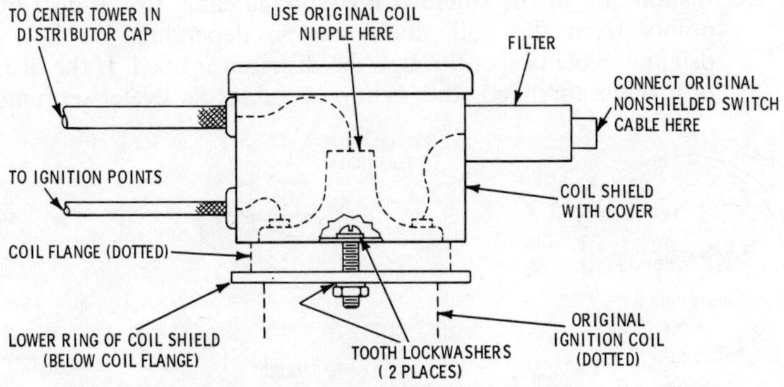

Fig. 7-30. Ignition-coil shield wiring.

cap is in alignment with the mating notch in the distributor bowl.

21. Attach the grounding cable to a nearby stud or bolt on the engine using a lockwasher from the kit. Allow sufficient slack in the cable so that the distributor cap can be easily removed for routine servicing.

22. Install the shielded coil on the engine, using tooth lockwashers from the kit. The original nonshielded cable from the switch connects to the outer terminal of the coil filter. Sufficient clearance must exist around this terminal to that there will be no shorts during engine operation.

23. Route the shielded spark-plug cables to their respective cylinder banks. Keep each group neat, and avoid excessive crossing over of cables. Snap each spark-plug shield over the proper spark plug, making sure the skirt of the shield fully covers the hex area of the plug.

24. Maximum shielding effectiveness requires that the shielded spark-plug cables of V-8 engines are grounded at or near their midpoint. Attach a large circular clamp to an existing cable support bracket on the engine as near the midpoint of the cables as possible. When the original cable support is plastic covered, remove the plastic in the area of contact so that a good electrical ground will exist when the circular clamp is attached. Route the shielded spark-plug cables through the clamp.

25. A few engines use air cleaners which may be sufficiently low to interfere with the shielded distributor cap. Should such interference occur, bend the air cleaner if possible or use one of smaller diameter. Small, high-performance units are used by many police vehicles and similar autos.

26. Installation of the shielded low-voltage cable to the distributor
points from the coil shield varies, depending on how the
original cable passes through the distributor bowl. If the distributor has a threaded stud, crimp or solder an eyelet terminal to

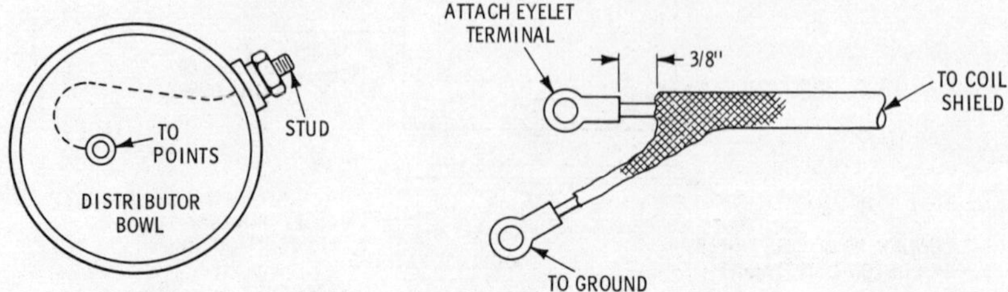

Fig. 7-31. Terminating shielded low-voltage cable to a stud.

the insulated cable and attach it to the stud (Fig. 7-31). If the
low-voltage cable passes through a grommet on the side of
the distributor, the portion of the flexible cable that is inside
the distributor bowl must be retained. (This flexible cable is
necessary for proper functioning of the automatic spark-ad-
vancing mechanism.) Splice and solder the small shielded cable
to the external portion of the flexible cable from the points
(Fig. 7-32). If the splice is made some distance from the distrib-

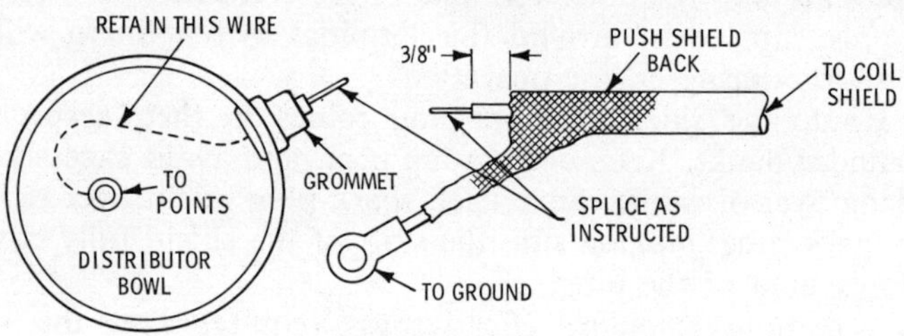

Fig. 7-32. Terminating shielded low-voltage cable to a wire.

utor, slide the shielding braid back and trim a length from
the inner wire only, before splicing and soldering the cables.
Connect the two conductors in line—without twisting them
together—and tightly tape the spliced area.

27. Connect the shielding pigtail terminal with tooth lockwashers
to a convenient grounded screw. Slide the shielding braid
toward the terminal until approximately 3/8 inch of wire is
exposed. Tape the shielding braid in this location to prevent
slippage during operation of the engine.

28. Reconnect the cable to the battery to complete the installation.

Johnson Shielding Kit

A shielding kit known as *Eliminoise* is marketed by the E.F. Johnson Company. It includes partial shielding for the ignition coil, the distributor, and the spark plugs, and full shielding for the ignition wiring to the spark plugs. None of the components are assembled. Because of the do-it-yourself nature of this kit, it can be adapted to almost any American-made automobile engine. In its present form it is not suitable for most marine engines. Fig. 7-33 shows the components of the kit.

In making an installation it is important to examine the original wiring arrangement at the distributor and to plan how the various parts are to be modified. It is desirable to use the most direct routing possible for all wires and cables without interfering with other engine components.

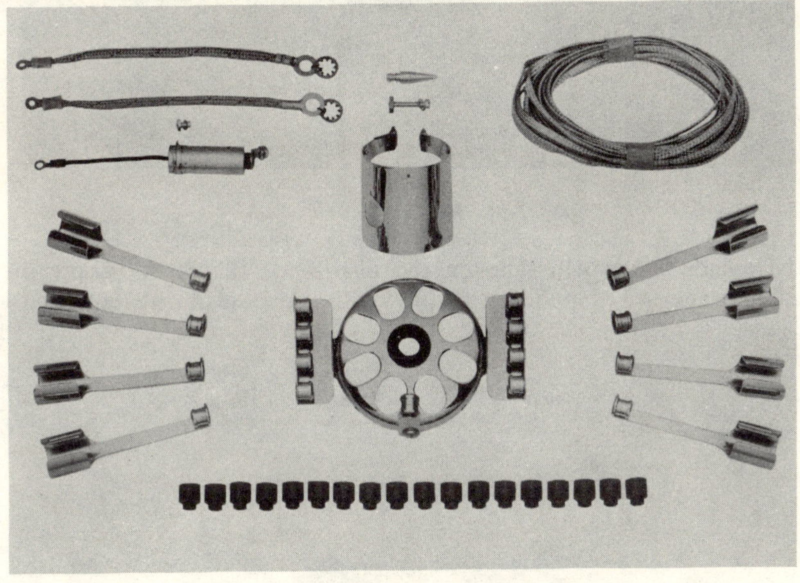

Fig. 7-33. *Eliminoise* **shielding kit.**

Fig. 7-34 shows a typical engine before shielding has been installed. Note that the wiring falls into a definite plan of distribution. The assembled shielding for the same engine is shown in Fig. 7-35. It follows the same general pattern as the original wiring. No two autos will be exactly the same; with a little study the best arrangement for a given engine can be worked out.

Be sure to make a sketch of the layout similar to the one shown in Fig. 7-36. Mark the type of cable brackets that will be used, and

Fig. 7-34. Typical ignition wiring.

the location they will take on the distributor shield. A variety of
cable brackets is supplied with the kit in order to fit different situa-

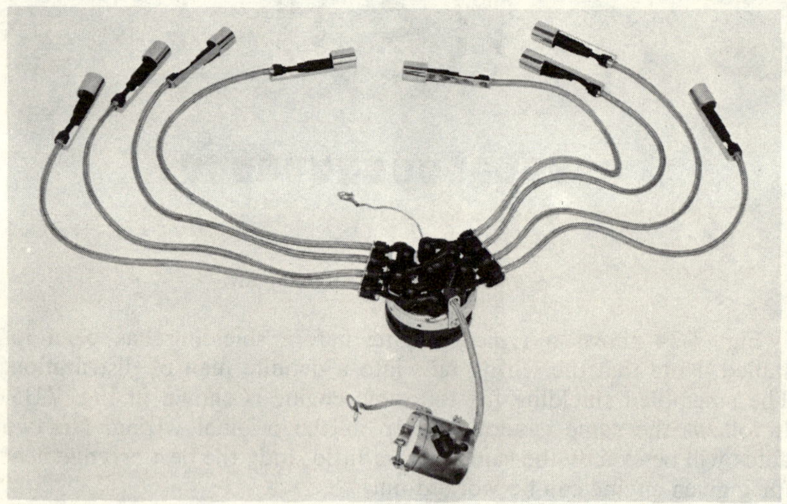

Fig. 7-35. Assembled shielding kit.

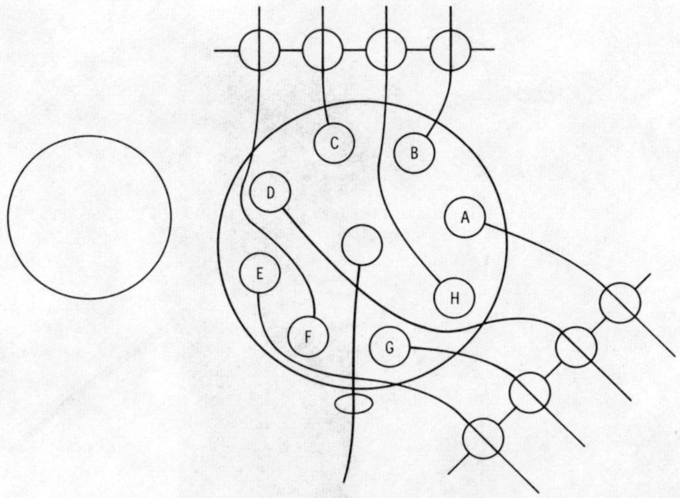

Fig. 7-36. Typical distributor-wiring sketch.

tions. Also indicate on the sketch which cable bracket hole each spark-plug wire will pass through. Use the shortest route for the shortest cables.

To install the kit, proceed as follows:

1. Disconnect one cable from the battery, preferably the ground cable. This will eliminate the possibility of damaging anything by electrical short circuits.
2. Remove the spark-plug cables from your engine. Use the letter tags, normally supplied with the kit, to mark each distributor tower, cable, and spark plug as the cables are removed.
3. Test fit the distributor shield to which the cable brackets and the bonding strap have been assembled according to your plan. The distributor bond strap must reach a convenient screw on the engine and still allow enough slack to remove the distributor cap for service.
4. Assemble the filter condenser to the coil shield. Position the shield on the top lip of the ignition coil. Rotate the shield for the best position of the condenser and the easiest access to the center tower of the distributor for the coil cable.
5. The coil has two small wires connected to it, one from the ignition switch and one from the distributor. Disconnect the one from the switch and connect it to the screw terminal on the filter condenser. Connect the short condenser wire to the coil terminal (Fig. 7-37). Connect the bond strap to the coil-bracket screw or other convenient point.

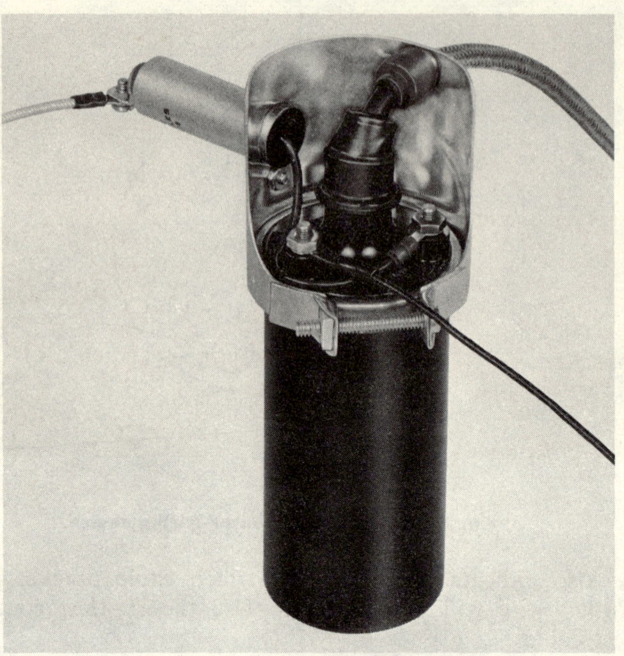

Fig. 7-37. Partial ignition-coil shield installation.

The remainder of the assembly can be made at a work bench. Be sure to maintain the shield in the proper position on the distributor cap while further work is being done. Before proceeding, examine the spark-plug cables for the following:

1. They must be of the radio resistance type. This has been supplied as original equipment on nearly all American-made automobiles since 1960.
2. The spark-plug end of the #1 cable must not be punctured. It is common practice of some mechanics when timing the engine to pierce this cable. If this has happened, or if there is any other evidence of deterioration, the wire must be replaced.
3. If wiring is over two years old, it should be replaced. The manufacturer of the shielding kit recommends that you purchase original-equipment wiring from your automotive dealer.

Modify the spark-plug wires as follows:

1. One end boot must be removed from each spark-plug wire, preferably the one on the spark-plug end. The boot removed

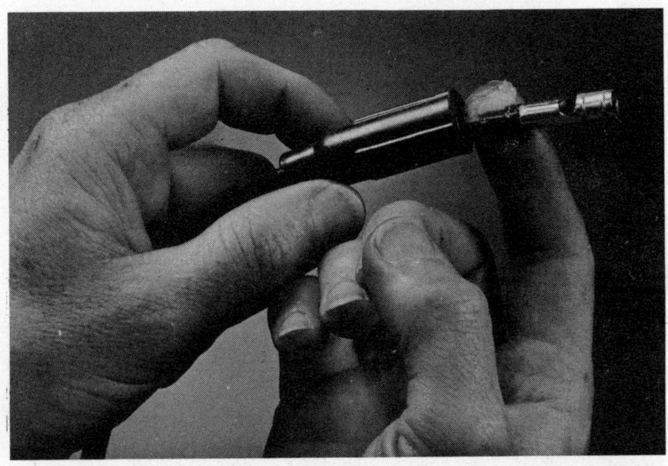

Fig. 7-38. Lubricating rubber boots on ignition wiring.

will depend on the automobile make since some manufacturers mold one end permanently. Lubricate the boot with petroleum jelly supplied to prevent damaging or breaking the electrical connection (Fig. 7-38).

2. Lubricate the small rubber grommets supplied with the kit, and slide one grommet to the far end of each spark-plug wire small end first.

3. Insert the wire either in the distributor shield or the spark-plug shield, depending on which of the original boots has been removed (Fig. 7-39).

4. Measure each wire from end to end. If the wires have angled ends, measure only the straight portion. Cut braid for each wire according to Table 7-1. Slide the braid on the wire. If you start from the spark-plug end, use the metal bullet supplied with the kit. If you start from the distributor end, the bullet is usually not required. If you find it easier, the braid can be opened before it is placed on the wire by passing an ordinary wooden pencil through, point end first.

Fig. 7-39. Installing shielding braid—step one.

Table 7-1. Braid length.

Wire Length (inches)	Braid Length (inches)	Wire Length (inches)	Braid Length (inches)
6	4	27	32
7	5	28	33
8	6	29	34
9	8	30	36
10	9	31	37
11	10	32	38
12	12	33	40
13	13	34	41
14	14	35	42
15	16	36	44
16	17	37	45
17	18	38	46
18	20	39	48
19	21	40	49
20	22	41	51
21	24	42	52
22	25	43	53
23	26	44	54
24	28	45	56
25	29	46	57
26	30	47	58

5. Make up the braid connection by fanning the braid out approximately 3/8 inch on the opposite side of the bracket before sliding the grommet into place (Fig. 7-40). Continue this procedure until all wires are made up at one end. Connect the opposite ends in the same manner. Replace the original rubber boots.

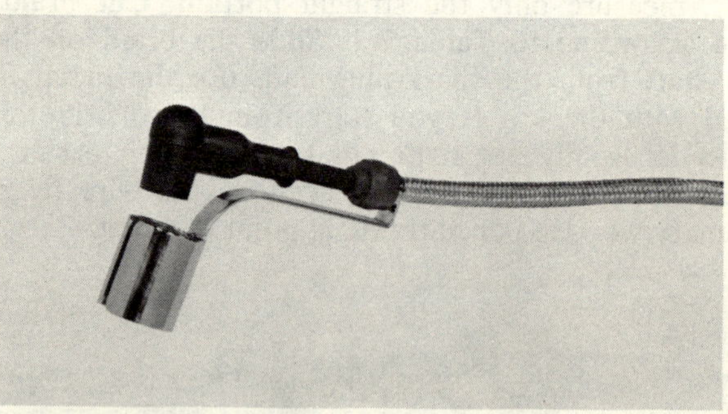

Fig. 7-40. Installing shielding braid—step two.

6. Insert the wires into the distributor cap. Use the original sketch to help in arranging the wires. They must be placed in the

proper rotation and distributed from the cap in the correct direction.

7. The spark-plug shields can be formed to match any of the angle fittings (Fig. 7-41). It is best to form these in conjunction with a spark plug to get the proper spacing with the wire connected.

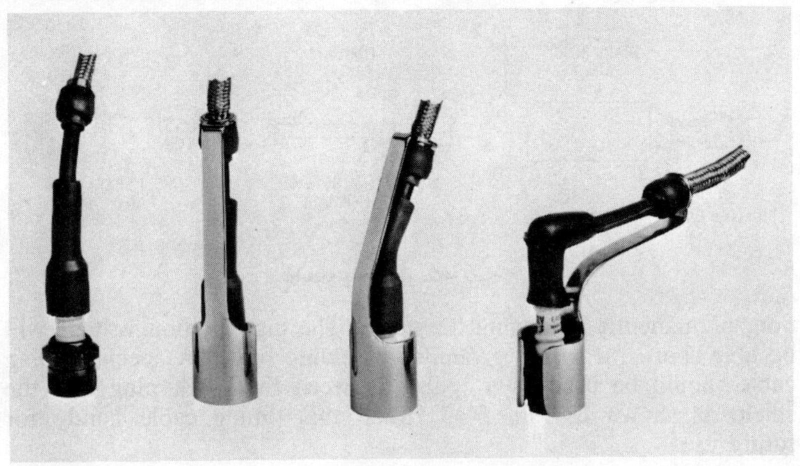

Fig. 7-41. Typical spark-plug shields.

Return the harness assembly to the engine, and install the distributor cap. Connect the spark-plug wires, using the letter tags for identification. When properly made up, the large ring end of the spark-plug shield will fit over the metal hex portion of the spark plug. This should fit firmly. If, when servicing the engine, it is found to be loose, it can be tightened by closing it slightly with a pair of pliers. A good ground connection at this point is absolutely necessary for satisfactory shielding.

To complete the installation, spark-plug wires are inserted in the original support brackets. Install the center-tower coil wire with the braid in the same manner as the spark-plug wires. Connect the ground strap from the distributor to any convenient ground point. Reconnect the battery, start the engine, and check performance.

MAINTENANCE

Continue to maintain the shielded ignition system as specified by the engine manufacturer. Follow the recommended intervals for cleaning and gapping spark plugs, replacing ignition points, etc. When mechanical work is done on the engine, be sure that all shielding connections are replaced and that the installation is not altered. The shielded spark-plug cables should not have to be removed from the shield unless they are damaged or the distributor cap is being replaced. A light amount of silicone grease (type DC-4) may be used as a lubricant when replacing any shielded cables in the cap.

When timing a shielded engine, do not pierce the shielded cable with a timing-light probe! This will puncture the high-voltage insula-

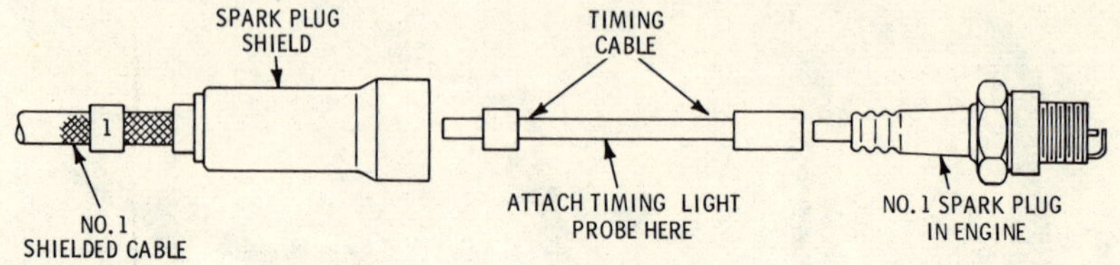

Fig. 7–42. Timing cable.

tion, permanently damaging the cable. The high-ignition voltage will be able short out to the grounded shielding braid. A special timing cable should be used as a spacer between the spark plug and the shield as shown in Fig. 7-42. Keep this timing cable handy for future use.

Index

112